COMPANY MANAGEMENT AND CAPITAL MARKET DEVELOPMENT IN THE TRANSITION

LICOS Studies on the Transitions in Central and Eastern Europe

1   Policy and institutional reform in Central European agriculture

2   Company management and capital market development in the transition

3   Marketization, demonopolization, and the development of internationally competitive enterprises in non-agricultural sectors of Central and Eastern Europe (forthcoming)

Edited at the Leuven Institute for
Central and East European Studies (LICOS)
Katholieke Universiteit Leuven
Blijde Inkomststraat 5
B-3000 Leuven, Belgium

Series editor: Marvin Jackson, Director LICOS
Assistant editor: Amy Crawshaw
Production assistant: Hilde Van Dyck

The research on which this publication is based was made possible by a grant from the Commission of the European Communities under its ACE program (DG 12).

Company Management and Capital Market Development in the Transition

Edited by

MARVIN JACKSON
VALENTIJN BILSEN

Avebury

Aldershot · Brookfield USA · Hong Kong · Singapore · Sydney

Published by
Avebury
Ashgate Publishing Limited
Gower House
Croft Road
Aldershot
Hants GU11 3HR
England

Ashgate Publishing Company
Old Post Road
Brookfield
Vermont 05036
USA

British Library Cataloguing in Publication Data

Company Management and Capital Market Development
in the Transition – (LICOS Studies on the Transitions in
Central & Eastern Europe; Vol. 2)
    I. Jackson, Marvin R. II. Bilsen, Valentijn III. Series
    658.152
ISBN 1-85628-998-2

Library of Congress Cataloging-in-Publication Data

Company management and capital market development
in the transition / Marvin Jackson and Valentijn Bilsen, editors
      p. cm. – (LICOS studies on the transitions in Central
and Eastern Europe; v. 2)
    ISBN 1-85628-998-2: £35.00 ($59.95 US: est.)
    1. Capital market – Europe, Eastern. 2. Europe, Eastern –
Economic conditions – 1989– . 3. Privatization – Europe,
Eastern. 4. Industrial management – Europe, Eastern.
    I. Jackson, Marvin R., 1932– . II. Bilsen, Valentijn, 1963– .
    III. Series.
    HG5430.7.A3C66 1994   94-35599
    338.94–dc20   CIP

Typeset by
Hilde Van Dyck
LICOS
Blijde Inkomststraat 5
B-3000 Leuven, Belgium

Printed and bound by Athenæum Press Ltd.,
Gateshead, Tyne & Wear.

Contents

Tables and figures

Contributors

Dr Istvan Abel is Professor at the Budapest University of Economics and Deputy CEO at the Budapest Bank.

Valentijn Bilsen is a research assistant at the Leuven Institute for Central and East European Studies, Katholieke Universiteit Leuven, Belgium.

Dr Alena Buchtíková is a researcher at the Institute of Economics of the Czech National Bank, Prague.

Dr Ales Capek is a researcher at the Institute of Economics of the Czech National Bank, Prague.

Irina Dumitriu is head of the Training Department of the Chamber of Commerce and Industry of Romania in Bucharest and Director General of the Romanian Business School Foundation.

András Giday is Managing Director of the Institute for Privatization Studies of the Hungarian Property Foundation in Budapest.

Ing. Miroslav Hrncír is head of Department at the Institute of Economics of the Czech National Bank, Prague.

Dr Marvin Jackson is Director of the Leuven Institute for Central and East European Studies and Professor in the Department of Economics, Katholieke Universiteit Leuven, Belgium.

Dr Jan Mujzel is Professor at the Institute of Economic Science of the Polish Academy of Sciences, Warsaw.

Teodor Nicolaescu is Director General of the National Agency for Privatization, Bucharest.

Dr Ognian Panov is Professor at the Center for Market Economy and Entrepreneurship, Sofia.

Kristofer Prander is a researcher at the Foretagsekonomiska Institutionen, Stockholm University, Sweden.

Dr Agnes Sári-Simkó is the Director of the Legal Affairs Department of the Institute for Privatization Studies of the Hungarian Property Foundation in Budapest.

Dr Stavros Thomadakis is Professor at the University of Athens, Center of Financial Studies, and the City University of New York, Baruch College, Department of Economics and Finance.

Foreword

The East European economic transition process has been going on for more than five years now. The economics of this transition seem to resemble more and more the famous 'globe of knowledge' defined by Pascal. As the 17th century French philosopher put it, the bigger this globe is, the more it is in touch with its environment, i.e. the lack of knowledge.

The ACE program of the European Union (former European Communities) has done very much to make this globe grow which means it did not only help carry out research but it also stimulated it by opening up new perspectives for science. It is namely also due to this program, an element of the PHARE scheme of assistance to the transition economies of Eastern Europe, that researchers venture into formerly unexplored fields of scientific knowledge related to the economics of the transition.

The microeconomics of the transition is such a field. Since the systemic and legal change has been accomplished or is coming to an end in most countries of the region, the transformation of enterprises into real actors of a market economy became one of the key issues of economic research related to the transition.

The ACE network organized by LICOS to explore the linkage between company management and capital markets in the transition consisted of research institutions in three West and five East European countries. Each of these institutes has a renowned expertise in management and corporate finance related research in general and concerning their respective countries as well. The participants of the network met on two occasions: a starting workshop was held in Leuven, Belgium in June 1992, and results were discussed and evaluated in Prague, Czech Republic in April 1993.

This network had to set broader and more ambitious research objectives than just the analysis of the aforementioned linkage between management and capital markets. It had to identify the elements of 'mainstream' industrial organization theory that could become helpful tools in the analysis of the behavior of those enterprises in the transition economies which are facing the problem of inefficient financial intermediation. In fact, it had to put together the elements of a transition-related definition of the firm. This is by far not a simple task, and could not be solved by this alone. This enterprise related analysis was also intended to shed light on the reasons for the very high exposure of many East European firms to capital markets.

Another important task of the network was to present the development of capital markets in the East European countries selected for the project. This was necessary in order to explain why many enterprises, albeit strongly exposed to domestic capital markets in many cases, cannot find adequate financing on these markets. They are therefore de facto obliged to request involuntary help from their suppliers to get at least temporary liquidity by not paying bills to them.

This necessity is at the core of the by now famous 'inter-enterprise debt', or the similar 'queuing' problem. It is very interesting to see from several country studies of this volume that 'queuing' does not necessarily disappear if a new, more drastic bankruptcy legislation is introduced. It simply puts more burdens on the commercial banking system while it does not really offer any reasonable solution to the problem of persistent financial imbalances in the enterprise sphere.

The results of the research published in this volume should and, hopefully, can find their way to at least three well definable sorts of professional readership. The first one is researchers and practitioners of privatization. The second one is researchers of corporate finance, financial policies and markets, and bankers. The third audience for these studies should be economic policymakers or, with a perhaps better sounding expression, the 'architects of transition'. Let us briefly summarize what these professional audiences could expect from the book.

The studies offer a deep insight into policies and practices of privatization strongly differing from each other across transition economies. Several privatization techniques using capital market instruments would eventually make it possible for state firms to obtain equity financing from the capital market. This would be in conformity with current practices of enterprise financing in most OECD countries where credit has a relatively minor role as compared to equity financing. Capital markets are, though, quite underdeveloped even in the most progressed transition economies.

This leads to a second important issue extensively analyzed in several studies of the book. This is inter-enterprise debt with more than one important implication for enterprises and government policies. It is evident that these 'techniques' of involuntary financing of one firm by another are, to a certain extent, substitutes for a well functioning capital market. Would it follow from this logic that the creation of a more or less efficient and well capitalized capital market could gradually eliminate 'queuing'?

The answer is made very difficult by the fact that, as the research points out, debt claims cannot be exactly compared with assets. The reason is a special phenomenon called the 'fuzziness' of debt claims. There might be serious differences between the official or contract-based value, the still more or less theoretical discounted value (as the one used in debt-equity swaps) and the ad hoc market value of debt claims. It is a very interesting topic of contract economics that the market value of debts, first of all smaller inter-enterprise debts, emerges in the precise but volatile conditions of a given action of debt settlement and could change very much from one day to another.

Capital markets could, eventually, not take the described direction of development, and remain undercapitalized and institutionally backward. This scenario would still not cut off the road for transition economies toward developing a modern financial system. This could then be bank-based instead of capital market based with a high degree of 'governance' control of banks over enterprises. The studies prepared by the members of the research network also provide ample food for thought for any expert wishing to assess the likelihood of such a financial market development in Eastern Europe.

This collection of high quality and up-to-date studies was prepared by a network of scholars with a very good knowledge of the field. Their work was coordinated by the director of LICOS, Professor Marvin Jackson, one of the most outstanding international experts of the microeconomics of the East European transition. His editorial work has further increased the quality of this book.

The knowledge offered by this book is very valuable to the scientist, but it can give the Western businessman doing trade with or considering investment in Eastern Europe much intellectual support as well. The book is warmly recommended to any economist trying to better understand the interaction between enterprises and financial systems which are both trying to adjust to the rules of the game of the fast emerging market economy.

Adam Török
Director, Research Institute of Industrial Economics
of the Hungarian Academy of Sciences, Budapest
Visiting Professor, LICOS, Leuven, Belgium

Preface

The aim of this book is to provide a collection of papers that describes and analyzes aspects of company management and the development of capital markets in the transition economies of Central Europe. It is needless to say that firm management and the development of capital markets are key elements in the evolution towards a fully established market economy. They are closely interrelated with each other in the field of company governance. The limitations and possibilities of company governance structures determine the degree to which the new market economies are able to allocate scarce resources into the most productive investments. Hence, it codetermines the production and income growth of these economies and as a consequence the welfare of the people living in the transition economies.

In the move from a centrally planned economy to a market type of economy, the reallocation of production from state into private hands is one of the urgent steps that inevitably has to be taken. The first two chapters in this volume deal with the issue of privatization for transition economies in general. Marvin Jackson discusses the role of privatization and the development of ownership institutions for the governance of companies in a transition economy. The necessary theoretical arguments are presented to investigate the importance of ownership changes for company governance. Valentijn Bilsen presents a comparative study of privatization methods in the Czech Republic, Hungary, Poland and Slovakia, and explores the relation between privatization, company management and performance. The subsequent chapters discuss issues of company management and capital market development for each country separately. These chapters are written by researchers from the Central European countries who present country specific information and analyze the relevant problems. Miroslav Hrncír explains the properties of financial intermediation in the Czech Republic and its implication for company governance. Alena Buchtíková and Ales Capek focus on the financial structure of the Czech enterprises and discuss the future role of the banks. Jan Mujzel discusses the changes in the company management of state-owned enterprises in Poland during as well as before transition, and investigates its determinants. Irina Dumitriu and Teodor Nicolaescu clarify the factors that caused the slowdown in the privatization process and of the formation of a capital market in Romania. Ognian Panov analyzes the privatization effort for Bulgaria and discusses the development of its young capital market. Istvan Abel and Kristofer Prander analyze the current condition of the Hungarian capital market and delineate how its constraints lead to a limited access of funds needed for restructuring by the enterprise sector. András Giday and Agnes Sári-Simkó describe the interaction between privatization and the development of the capital market in Hungary. In the concluding chapter of the book, Stavros Thomadakis provides some common features of transition economies. Also the lessons concerning financial system design and corporate governance that can be drawn on the base of the experience of the Central European countries so far are presented.

This book results from a research network on 'Company Management and Capital Market Development in the Transition'. It was sponsored by the ACE program of the European Community. The coordination was done by the Leuven Institute for Central and East European Studies (LICOS) of the Katholieke Universiteit Leuven, Belgium. The aim of the network was to bring specialists from Central, Eastern, Southern and Western Europe together to study the governance and management of enterprises during transition and to investigate the gradually growing market environment which is needed for an efficient operation of firms. Starting from the heritage of socialism, that is marked by its shown inability to effectively allocate capital, the countries under study are developing their own way towards a market economy. The size of this move is of an unprecedented scale where the joint aspects are often hidden under the diversity of problems, government policies and reactions of the new born private sector. For instance, every country that was studied has its own 'success formula' to privatize large enterprises. In the Czech Republic and Slovakia this was the voucher scheme. In Hungary the method of self-privatization played an important role. To unravel a fundamental part of this phenomenon (transition) was a major challenge for all of the researchers involved in the network.

A necessary ingredient for success was the combination of the profound knowledge and expertise of leading economists coming from the transition countries, and the experience and insights from their colleagues from the established Western European market economies. On the one hand, this resulted in the supply of detailed and country specific information which would otherwise have been the privileged knowledge of a small group of insiders. On the other hand, the insights based on the analysis of full grown market economies served as a benchmark by which the relevant phenomena in the transition economies could be detected and assessed. This book can be considered as the materialized result of a fruitful cooperation between all the researchers of the network, who tried to have an eye for both the general aspects of the research topic as well as for the relevant details leading to a particular outcome.

Marvin Jackson
Valentijn Bilsen
Leuven, July 1994

Acknowledgements

The original idea of bringing researchers from West and East Europe together in a research network to pool information and knowledge on developments and analytical methods was developed at the Leuven Institute for Central and East European Studies. Highly acknowledged help in this project was provided by Jan Klacek (Institute of Economics, Czech National Bank) and Stavros Thomadakis (University of Athens, City University of New York).

The network comprised two workshops, one in Leuven on 12-13 June 1992 and one in Prague on 24-26 April 1993, and some bilateral visits among the researchers involved. The first version of the papers was presented in Prague. The LICOS administration and the team under the direction of Jan Klacek did a splendid job in organizing respectively the first and the second workshops.

The financial support from the ACE program of the Commission of the European Communities, by which most expenditures of the project were funded, is highly acknowledged. The research and editing of Valentijn Bilsen was financed by the Belgian National Science Foundation (Nationaal Fonds voor Wetenschappelijk Onderzoek).

Katrien Verhelst provided excellent and diplomatic skills to keep the network outlays within the hard budget constraint. Hilde Van Dyck provided effective logistic support, including the typesetting of the whole book, with a highly appreciated good cheer. And last but not least, the professional English language editing and coordination with the publishers of Amy Crawshaw, characterized by a never waning motivation, was a crucial ingredient in the realization of this book.

Marvin Jackson
Valentijn Bilsen
Leuven, July 1994

xix

Abbreviations

AGs	Akteingesellschaften (public limited companies)
BCC	Bank Consolidation Company
BCD	bank credit/total liabilities
BFTB	Bulgarian Foreign Trade Bank
BIB	Bulgarian Investment Bank
bn.	billion
BNB	Bulgarian National Bank
BSP	Bulgarian Socialist Party
CEC	Central European country
CEE	Central and Eastern Europe
CK	Czech crowns
Comecon	Council for Mutual Economic Assistance
CPI	consumer price index
CR	cash ratio
CSK	Czechoslovak crowns
CSO	Central Statistical Office
CUR	current ratio
DA	total liabilities/assets
EME	emerging market economy
ESOP	see MRP
EU	European Union
FDI	foreign direct investment
FPB	First Private Bank
GDP	gross domestic product

GmbH	Gesellschaft mit beschränkter Haftung
HUF	Hungarian forint
IBRD	International Bank for Reconstruction and Development
IMF	International Monetary Fund
IPF	Investment Privatization Fund
IPO	initial public offering
JV	joint venture
LDA	long-term debts/assets
LDE	long-term debts/equity
MBO	management buy-out
mln.	million
MoP	Ministry of Privatization
MPP	mass privatization program
MRF	Movement for Rights and Freedom
MRP (ESOP)	Employee Share Ownership Program
NAP	National Agency for Privatization
NICs	new industrialized countries
NIF	National Investment Funds
NJF	National Investment Fund (Poland)
OPZZ	All-Polish Trade Union Alliance
p.a.	per annum
PA	profits before taxes/total assets
PBDS	primary bad debts/monthly sales
PE	profits before taxes/equity
PHARE	Poland and Hungary Action for Restructuring of the Economy
PI	profits before taxes/interest payable
POF	Private Ownership Fund
PPI	producer price index
QR	quick ratio
R&D	research and development
SA	sales/total assets
SII	State Insurance Institute
SMEs	small and medium enterprises
SOE	state-owned enterprise
SOF	State Ownership Fund
SPA	State Property Agency

SSB	State Saving Bank
TB	Treasury bill
TBDA	total bad debts/assets
TBDD	total bad debts/total debts
TBDE	total bad debts/equity
TBDS	total bad debts/monthly sales
UBB	United Bulgarian Bank
UDF	Union of Democratic Forces
UND	Union for New Democracy
USD	United States dollar
USSR	Union of Soviet Socialist Republics
VAT	value added tax

1   Property rights, company organization and governance in the transition

Marvin Jackson

Introduction

The purpose of this chapter is to explore one of the two key processes in the transition of the Central and Eastern European countries from communist rule to western-style democracies with market economies, changing the property rights system or privatization as it has become popularly known. We shall first consider a number of technical issues before turning to a discussion of the institutions of organization and ownership. After laying this foundation we shall turn to the problems of the transition itself.

We shall consider how the agents in the transition countries have approached their own privatization processes. There appears to be considerable diversity, both in the motives for privatization and the specific means selected to go about it. At the end of the chapter, we shall try to generalize what we have found to identify some lessons for institutional change and policy.

Some technical issues in changing property rights

Privatization and the means to do it

For purposes of clear definition, privatization means the reassigning of three basic property rights in physical and intellectual assets - control, income and alienation - to physical and legal persons other than the state, as provided in the normal commercial code. This includes simple personal proprietorship, partnerships, cooperatives, as well as limited liability and joint-stock companies. This chapter focuses on the privatization of the assets of state-owned enterprises (SOEs). Nevertheless, other forms such

1

as restitution, sale, or leases of shops and individual assets have been important.

In addition, of course, much new private property is created when private investors, domestic and foreign, decide to create completely new private assets. What is different, of course, in these cases, is that there is less change of ownership in the sense of what happens when a former state-owned company is sold to investors or given to its employees. Only the latter is what we would refer to if we were talking about privatization in a market economy.

In transition countries, one of the first steps of the privatization process is to end the legal existence of the SOEs, either through liquidation or commercialization:

1   Liquidation means transferring the assets of the SOE to other organizations and persons (legal and physical) and terminating the legal identity of the SOE (taking it off the company registration book).

2   Commercialization, sometimes called conversion or other names, means legally reorganizing the SOE into one of the normal forms of business organization permitted by a country's commercial code, usually as a joint-stock company or a limited liability company, without transferable shares.

As in all practical aspects, transition countries have differed in the extent to which SOEs have been converted into other legal forms prior to privatization. In the case of Czechoslovakia, SOEs were converted directly into new organizational forms at the time of privatization, without intervening commercialization. Other countries have differed in the extent to which they have used mass commercialization or case by case commercialization. Also, there have been differences in the extent to which simultaneous commercialization and restructuring have been undertaken.

Restructuring has referred to several different features. It can be financial restructuring or changing the values of actives and passives on the company's accounting records.[1] It can refer to organizational restructuring or changing the partitioning of activities (including closing departments or liquidating subordinated units and divisions), changing the structure of management, and changing personnel. It can also refer to technological restructuring such as modernizing equipment, changing process technologies, and others.

Methods of transferring asset ownership: ownership changes can take the form of transferring individual physical assets (such as machines, buildings, supplies of materials and finished goods, lorries, etc.), shares of joint-stock companies and other forms of company ownership, and whole companies.

The transfer can involve several different types of compensation, such as:

1   ordinary sale for cash;

2   sale for credit;

3   lease with right to purchase;

4   sale at a discount price;

5   free transfer.

The transfer of assets can be arranged in different ways:

1   restitution to the owners prior to communist nationalization;[2]

2   priority offer to the employees and/or managers without special terms;

3   priority offer to the employees and/or managers with special terms, such as free transfers, discount prices, or subsidized credits, so-called leveraged buy-outs;

4   sales on open auctions;

5   sales on stock exchanges;

6   negotiated sales to special or strategic investors;

7   share transfers to social organizations, such as pension funds;

8   share transfers to the public as a whole or through the intermediary of so-called voucher programs.

Various privatization programs have been developed both for individual privatization, that is one company at a time, or forms of batch privatization or mass privatization. In the latter cases, several companies are prepared and offered as a group.

Free or price discounted shares are usually limited to the employees of a given company. The more spectacular schemes are those involving the distribution to all citizens of vouchers or coupons, for free or for a nominal price, that can be used to buy assets or shares in companies being privatized. Such vouchers can be directly exchanged for shares or indirectly invested through the intermediary of special investment funds. Other forms of transfer with subsidy are through credit purchases at special interest rates or lease-purchase on special terms.

Some approaches to the analysis of privatization

Perhaps three quite distinctive approaches to the analysis of privatization and changing property rights can be identified:

1   the political economy and the process of privatization - analyze the processes that are involved in order to understand the behavior of the various participants or agents who are involved in it;

2   the effectiveness or speed of privatization - investigate the factors that explain the results of the behavior of agents in the privatization process in order to understand the speed and effectiveness of it;

3   the effects of the achieved forms of privatization on the behavior of agents in enterprises - take the processes that are observed for granted and inquire into the effects of the resulting changes in property rights on the behavior of enterprises and the agents, both in the enterprises and in the agencies that might deal with enterprises, the policy bodies of the governments.

Political economy and the process of privatization. Privatization should reflect the behaviors of many different agents, including:

1   politicians who must pass legislation;

2   government officials who pass regulations and carry them out in privatization ministries and agencies;

3   enterprise managers;

4   workers and trade unions;

5   financial institutions;

6   various agents who wish to acquire companies and individual assets being privatized.

All the agents involved may be said to be acting in their own self interests. Politicians, of course, are commonly seen as pursuing some broader political criteria such as seeking through a redistribution of assets among owners arrangements that might be more equitable, more supportive of democratic power, or more efficient in society's interests. Politicians also serve special groups who support them and have their own personal wealth and income in mind. All of the other agents will have similar goals and see similar ways of going about achieving those goals.

One way of looking at all agents together is to consider privatization as a massive asset market. Some agents are offering assets and others are seeking to acquire them. The means of exchange are not only money, but also political support and influence. The agents' behaviors are affected by many different information problems and uncertainties both about the quality of the assets available for exchange, but also about the likely behaviors of other agents.

Effectiveness or speed of privatization. The economic criteria of privatization is initially rather simple. It is to put assets in the hands of those who value the assets most highly. It can be demonstrated that when assets are controlled by those individuals in the society who value them most highly, the assets will also generate the highest discounted present value of net returns from the society's point of view.

According to the Coase Theorem, if there was already a market economy in which there were no transaction costs, no matter what the initial distribution of assets was among the members of society, the result would be an efficient allocation of those assets from the point of view of control.[3] When one member had a higher valued use of a given asset than the member in initial possession, then the asset market would permit both to gain by a transaction of the first renting or buying the asset from the second. Either way, control passes to the one with the highest valued uses or who values the asset the most.

It is well known, of course, that no economy is free of transaction costs. In existing market economies, among other problems, there are no perfect asset markets. Because of this, privatization programs in existing market economies have to be based on more ad hoc criteria designed to reflect particular transaction costs structures.

Since the goal of transition is to achieve economic systems like those in existing market economies, privatization programs should be designed around criteria that are applicable in existing market economies. However, the transition state imposes its own special transactions costs. These have to do with:

1   the uncertain dimensions of the political commitment to privatization;

2   the stock of human capital and organizational routines; and

3   the imperfect structure of the political and institutional framework.

The stock of human capital and organizational routines reflect what people have learned and past organizational structures and behavior. The new political and economic systems are still in the process of perfection.

Effects of the achieved forms of privatization on the behavior of agents in enterprises. The effects on the behavior of agents in enterprises of new property relations achieved as a result of privatization is expected to result sometime in increased productivity and national income in the transition countries. Whether this happens or not depends upon many different possible parameters of the institutions that are being created. It is not self evident that any or all changes in property rights will do this.

Property institutions in market economies

The range of institutions found in Western Europe are taken to include the institutional goals of most persons in CEE. This section attempts to describe the main economic dimensions of firms, ownership, governance institutions and public ownership. These are not institutional facts, but taken rather in terms of the economic relations that are thought to be at work in the processes of institutional formation.

Conceptual issues in organization of firms

Definitions of the firm. The terms, 'the firm', 'the company' and 'the enterprise',[4] can be used interchangeably. There are several useful definitions of the firm found in the economic literature. Chandler, for example, gives four definitions: the neoclassical (the firm as a production function), the principal-agent (the firm as a nexus of contracts), the transactions cost (the firm as a hierarchy), and the evolutionary (the firm as a complex organization).[5] The latter three views of the firm are most useful in discussing the organization and size of the firm.

Two views of firms and markets. As already discussed in some detail there are two, sometimes conflicting but actually complementary, approaches to explain what is important about the nature of the firm and the boundaries between firms and markets. The first approach extends the concepts of neoclassical economics through an application of transactions costs in the theory of contracts to explain how markets work and when firms and other hierarchies might be more effective for organizing the social division of labor. The organization of firms is seen as a consequence of extending market competition. The second approach sees a market economy essentially as an opportunity for economic agents to create efficient organizations that compete not only through price-quantity offers, but by technological and organizational innovation. Firms compete through organizational evolution, not by discovering efficient equilibriums.

The firm as a response to opportunity: transactions costs and contract theory

In Ronald Coase's original article on why firms are used and not markets, the possibility of reducing transactions costs gives rise to formal business organization.[6] The boundary between firms and markets is set by the combination of institutions that minimizes the combined costs of transactions and production. In the subsequent development of contract

6

theory, a fine boundary between simple market transactions and transactions within a hierarchy has been blurred by the possibility of independent agents contracting for various long term relations that capture elements of costs reduction available through both markets and firms.[7] There remains, however, a theoretical division in the literature over the role of bounded rationality. The presentation that follows is that of the Williamson approach that makes bounded rationality a focus of why firms are organized.[8]

The essence of the firm might be seen as the typical employment relationship, in which the worker submits to detailed directions of activities in return for a wage and usually some security of employment. According to Williamson, the employment relationship and indeed the extent of all activities and real assets necessary to conduct employment, and therefore the dimensions of the firm itself, tend to be defined by three relationships:

1 the frequency of transactions between the two agents;

2 the amount of transaction specific investments required on the part of both parties; and

3 the difficulty of writing a complete contract between two agents.

Asset specificity, according to Williamson, can take three forms: site specificity, physical asset specificity, and human asset specificity.[9] The making of such transaction specific investments by both agents can result in large increases in productivity, but at the same time lock both sides into a relationship in which either is subjected to possible opportunism by the other. Incomplete contracting arises because of uncertainty, bounded rationality and imperfect information. It is impossible to foresee all events. Opportunism can be reduced by integrating the activities into one management unit under hierarchical authority and a long run contract.[10]

Promoting the employment relationship, in contrast to arm's length contracting, are such investments as specialized training and learning by doing for the company's peculiar assets and situation, improved communications from personalized knowledge and even the use of special languages, and institutional and personalized trust. The specific investments required of a good manager are especially great. Once such investments are made, both sides stand to lose: the employee loses because he cannot turn to other employment without loss of productive value; the employer loses because of the costs of disruption and training that replacement entails. While a single employer facing many employees might be tempted to threaten the many by making an example of just one, such an employer would risk paying more in the future because of the reputation gained.[11]

Employment is not the only relationship promoted by asset specificity. The efficiency of investments in capital might also be enhanced by joint ownership. For example, a power plant might be adopted to the specific characteristics of coal from one pit, while machinery could process the coal, say through a conveyor, so that it could be used efficiently by only one user, the power plant at the end of the conveyor.

In this view, benefits of employment and common ownership (integration) can be seen as resulting from the greater difficulty of writing contracts and enforcing them. If agents are prevented from creative application of contracts, then hierarchies might be overencouraged. Freedom to seek new ways to enforce contracts is also important.

Organization and the firm as a maker of markets

The focus on transactions and contracts as the basic unit of analysis contrasts to an approach which considers the firm and its organization as a unit of analysis. This, for example, is the choice of terms chosen by economic historian Alfred Chandler to differentiate his approach from that of Oliver Williamson.[12] Chandler's focus is on the organizational history of large American, British and German firms and an explanation of why some acquire the long term ability to grow and prosper in international competition.[13] His answer is that some firms, through investments in large-scale production, distribution channels and management structures, acquire first mover advantages in the form of organizational learning. His emphasis is on how the firm makes full use of production and distribution economics, not on reducing transactions costs. Organizational learning, something carried on independently of the individual managers, allows firms to define markets through function and strategic competition. Also, these forms of competition are more important for defining and changing market shares and profit than is classical price competition.

Chandler's complaint with neoclassical and principle-agent theory is that it "... neither deals with the firm's physical facilities, and human skills and the resulting revenues on which the current profitability and future health of the enterprise depend."[14]

Although Chandler is not overly concerned with theory, he sees his own work as empirically relevant to evolutionary economics as developed by Nelson and Winter, with its stress on organizational routines.[15] Chandler's central point is worth a rather long quotation:

> ... managerial skills were based on learning carried on within different levels of the hierarchy - the operating units, the functional departments, and, as the firm grew, the product and geographical divisions and, of course, the corporation offices. Such learning was a process of trial and error, feedback and evaluation. *It is more organizational than individual* [emphasis added]. Even the skills of individuals depended on the organizational setting in which they were developed and used. If these company-specific and industry-specific capabilities continued to be enhanced by constant learning about products, processes, customers,

8

suppliers and other workers and managers within the firm, the enterprise was usually able to remain competitive and profitable. If not, its market position deteriorated.[16]

Chandler's histories describe how in the 19th century, American firms in capital-intensive industries pioneered the U-form with its separation of strategic (with its functional staff departments) and operational management functions in order to take advantage of the newly available production and distribution technologies. He followed this with an identification of the M-form, multi-unit and often multi-national enterprises through which American firms dominated world markets until such organizational techniques were successfully adopted and even improved upon by the Europeans and the Japanese in the late 1960s and 1970s. And although Chandler's work reveals failures as well as successes among large companies, he is impressed with the way industry leaders, once established in the 20th century, have remained leaders. One of his predictions is worth repeating:

> I am willing to predict not only that the modern industrial firm ... will be as powerful an economic institution at the beginning of the 21st century as it is in the 20th, but that a number (though certainly not all) of the U.S. global leaders today will remain as dominant in their global industries in the future as they have been in the past. Moreover, their rivals will continue to be, as has been true in the past half-century, not entrepreneurial start-ups but comparable enterprises from overseas or from related industries.[17]

It should not be overlooked that Chandler's prediction applies to the group of capital-intensive industries, such as automobiles and electrical equipment. It does not apply to:

> ... labor-intensive industries where the economies of scale and scope provide smaller costs advantages and where a slower moving large enterprise can be encumbered with competitive disadvantages, that is, in such industries as textiles, apparel, furniture, lumber, leather, specialized materials, tools, instruments, and the like.[18]

Chandler himself has added very important dimensions to any correct view of the modern market economy and one that is necessary to understand the strategic behavior of large oligopolistic firms that play a critical role in it. His work has also provoked important studies of the Japanese firm.

Chandler's work has had a strong influence on views of what are the essential elements of success of the western capitalist economies. This is partly because he has pushed economic historians into the study of company organization. Also, he provides a linkage with modern management literature with its essentially normative concern with how to prevent a decline of large company efficiency.

At the same time, however, his work has been subjected to important qualifications and criticism. Two important issues are the role of market forces and the role of small enterprise.

Chandler is no proponent of central planning nor does he see cooperation among multinational enterprises as setting prices and allocating markets in a neo-Marxist or Galbraithian view. In his view, large markets were necessary, as was the prevention of price-output cartels by small firms, in order to provide the field of response and the markets into which successful firms could penetrate and grow. He would not appear to be an advocate of protected markets, especially in small economies. Also, in his view, British organizational capabilities failed to develop, among other reasons, because firms did not compete vigorously with each other.[19]

He has been criticized as having overestimated both economies of scale to the neglect of innovations and the role of company control over investments in expanding scale and scope as opposed to funds generated from the cost-reduction and market-creating effects of innovations.[20] He gives little importance to static comparative advantage and exhibits little concern for problems of possible monopolistic behavior of large firms or the possible problems arising from the separation of ownership and control. His emphasis on the role of managerial capitalism does not seem to adequately consider the societal context in which managers develop and are appointed to their places.[21]

German industrial historian Jurgen Kocka suggests that, while German firms engaged in various forms of cooperation, they also were seriously competitive, something that Chandler sometimes tends to underestimate.[22] In his opinion, Chandler also neglects the important contributions of small and medium enterprises, not only in the German case but generally.[23] Chandler admits this point, but only in the sense that the dynamics of interactions between large firms and SMEs should be studied. As the above quotation makes clear, he does not see entrepreneurial start-up as important in industries dominated by scale and scope economies. This contrasts with the current positive role in innovation and growth now being given to SMEs in the literature.[24] Also, his view is not shared by a number of other advocates of evolutionary economics, who would stress not only the organizational routines of those who succeeded, but also those growing up and sometimes failing to survive the evolutionary process of market competition.

In summary, Chandler provides a link to modern management literature and a point of view that matches the critical roles of strategic competition among large-scale enterprises in international markets. His analysis, nevertheless, should not be taken as ensuring that effective and growing economies automatically can be generated by creating large enterprises and staffing them with well trained managers.[25] Nor should it lead to an underestimation of conventional market competition and the role of comparative advantage in open international market competition. For example, a respected analyst of the Japanese firm stated that: "... the first key to an understanding of Japan's industrial performance is the ability of the firm to react quickly in response to changing market conditions ..." [26].

A main weakness in Chandler's approach is its inability to explain how small firms found the keys to grow into large firms, and what some large firms did that caused them to lose out in international competition, a point that he, himself, stresses as being the major challenge raised by his work.[27] Chandler shares a weakness with other evolutionists in leaving survival to a stochastic process, rather than trying to explain it in a positive theory.[28]

Ownership of firms

In this section, we shall make use of yet another definition, Putterman's ownership firm.[29] In his view, the firm itself is a legal person that can own property and contract. It is also an asset that can be owned by one or more real persons. Obviously the ownership firm is a smaller set of entities than the organizational firm, if for no other reason than the non-owner human agents in the firm cannot be owned.

The full owner of the firm has three general property rights over the asset: control, usufruct (income), and alienation. Control rights give the property right holder the right to determine how the property will be used, although this can readily be seen to involve a principle agent problem if the property right holder hires someone to manipulate the asset. Usufruct rights mean the right to receive any rents or profits earned from using the asset. Alienation rights mean the rights to sell, give away or destroy the asset.

It is a rare situation where a property right holder has an unqualified right. The owner of an automobile does not have the right to violate traffic rules or to drive negligently. Income rights are subject to taxation. In Europe, people cannot legally sell themselves into slavery.

Importance of the real owner

The importance of the real owner is the impact real ownership has on the way the real assets of the firm will be used. In the absence of certain conditions (to be discussed further), it is society's interest that real assets be used so as to maximize profits or the present discounted value of net future returns. That is, in such a state the assets would be allocated and used in their most valued uses.

The real owner of all three property rights, whose well-being is otherwise not dependent on the actions of the firm, will be interested in deriving the maximum income from the firm. Therefore, he will use control and alienation rights to assure that end. The right of alienation is very important to this end. For it permits the establishment in markets for firms (as assets) in which other potential owners can offer the existing owner a higher income if they are aware of better uses of the assets of the firm.

Problems of common owners

Alienation rights are also important for the case of common owners if they have different time preferences. Without alienation they would disagree on policies of using the firm's assets. The creation of transferable ownership shares could solve the problem. A policy maximizing the present value at the going rate of interest could be chosen and then one owner could compensate the other.

A more complicated problem arises if the common owners must undertake monitoring, of employees and other agents, and other efforts in order to assure that maximum profits are earned. In this case, the problem arises of sharing the burden of such effort and of shirking on the part of one, or some, of the common owners. If there are many common owners, each with an equal share, no one would have an incentive to undertake personally costly monitoring for a small share of the gain. This is a common public good problem and, of course, arises when the shares of stock in a joint-stock company are widely held.

One solution would be to have a dominant owner. He or she would have an incentive to undertake monitoring actions even if the other owners would not.

Separation of rights of control from rights of income and alienation

The monitoring problem arises in the case where significant control rights are delegated to a manager by the owners of income and alienation rights. Because of asymmetrical information, the manager might have the incentive to divert wealth of the owners to himself. This would be less of a danger if the value of diverted wealth was less than the reduction in firm profits (in this case, it would pay the manager to bargain for part of this difference in the two sums). The manager should also be concerned about the long run value of his reputation, which if questioned could cause his income to fall.

Incomplete contracts, asset specificity, and the employment relation

The owners of a firm and the manager would have incentives to engage in a long-term contract that provided protection for specific asset investments on the part of the manager, while attempting to encourage the manager to maximize the value of the owners' assets.

Organizational design and the assignment of residual rights of control

Even within the firm, there is benefit to the attempt of all agents to agree in advance on what is to be done. Nevertheless, it is impossible to pre-specify all actions. Some are left to be allocated as residual rights of control. In a well designed firm, the residual rights of control should be assigned to those who are most likely to choose the socially efficient course of action. Those are likely to be the agents whose assets values are most highly contingent on the course of action chosen.

This means that for those actions likely to affect the human assets values, the residual control rights should be assigned to the manager and other employees. For those actions likely to affect the value of site and physical capital, the residual control rights should be assigned to the owner (holder of alienation and income rights). In adapting this principle of Oliver Hart, Louis Putterman would add as very important assets the ability to bear risk and to supply financing services. In fact, Putterman goes on to argue that possession of risk bearing capacity may be the necessary condition for investing in transaction specific physical assets.

Why workers are not commonly owners

In the west, there is no law against workers being owners, but this is not common. There are a number of reasons why.

To begin with, workers might encounter the problems of common owners already mentioned. They would need to have alienation rights and a market for shares in order to resolve differences in time discount rates. Also, if a large number of workers owned their firm, there might be problems of shirking over the monitoring of management.

It would also be more difficult for workers to accept waiting for returns in time-consuming production or to bear risk, given their likely smaller holdings of savings. These latter functions are done more easily by relatively wealthy investors, who are more likely to be owners.

The worker also would bear a special problem of risk. Skilled workers, who have been with a firm for a number of years, accumulate specific human capital in that firm. This capital would be at risk if the firm were to fail. Therefore, adding the financial assets of investment in ownership to the specific capital investment would subject a worker to high risk. It would be more likely that financial savings would be better invested in a diversified portfolio.

There are costs to a worker who does not invest in his employing firm. One is that management probably has higher costs of monitoring non-owner workers than owner workers. Therefore, wages of the former ought to be discounted in competitive labor markets. Also, it is possible that workers have less satisfaction working for the profit of others. If both costs were high enough, it might still provide net benefits for a worker to own a share of his employing firm.

13

Laffont and Tirole suggest a simple taxonomy based on the contrast between external and internal control.[30] External control is the possibility of regulating the variables that link the firm with outsiders (regulation of prices, quality, product selection, regulation of entry, cost auditing, etc.) while internal control covers the firm's inputs, cost minimization processes, incentive schemes (which affect managerial inputs), employment, investment, etc.

Public firms are subject to both external and internal control by the government. Private regulated firms are subject to external government control and internal private control, while an unregulated private firm is subject to both private external and private internal control. Public control, in their system, refers to regulation of specific firms, not to regulations such as tariffs, anti-trust and other regulations applying to all firms.[31]

According to conventional views, public and private firms share a number of the same problems. They have boards of directors representing dispersed owners. Their boards are known to exert too little control, because they receive too little information from managers, and in both, takeovers, in the public firms by controlling the election of the board, are considered weak means of control.

There are also conventional views of their differences, although it has been shown (Williamson, Grossman and Hart) that such differences arise because of the difficulty of writing complete contracts (see above). A public firm that does not leave shares for private investors would suffer from the absence of the capital market helping to monitor its managers. However, reducing the amount of traded shares would increase their illiquidity, leading to garbled market information. Possibly an answer to this problem would be for the government to issue non-voting shares, thereby increasing market liquidity without threatening loss of ownership control.

Another problem arises from the less precise objectives of public control compared to that of profit making. In this case, contracts with management are even less complete and, therefore, give more room for opportunism to managers. At the same time, managers of public enterprises who invested their own human capital to cut costs and increase profits might find the profit simply taken and reallocated by the government, whereas there would be no reason for private owners to do this. In addition to these problems, public ownership is subject to soft budget constraints. Both public ownership and regulation are subjected to pressure from interest groups. On the other hand, the public firm provides a means to achieve social welfare and for centralized control.

Governance functions of financial institutions

Berle and Means pointed to a weakness of the capital allocation system resulting from what they called the separation of ownership and property.[32] Others have since seen it in terms of a more general phenomenon of principal-agent problems. When professional managers run public companies whose shares are widely distributed among many relatively uninformed owners, high transactions costs of organizing proxy collections results in the average stockholder taking no direct voice in the management of company assets and remaining only with the choice of selling their shares if the dividends and capital gains do not seem satisfactory. The professional managers, many of whom come from their company's ranks and own rather few shares, appear to be free of share-owner control. When combined with significant weakness in product market competition and monopoly power, managers acquire significant power to maximize their own wealth by their control over company assets. The problem could be compounded significantly through the organization of holding companies and through mergers, horizontal and vertical. There have been at least three different answers to this problem, each however with its own controversy.

Chandler's positive managerial capitalism

Unlike Berle and Means and contemporary advocates of principal-agent theory, Chandler contends that independent management dominated governance is a necessary condition for a prospering large multinational enterprise. Only when corporation policy is made by directors who are full time managerial employees will companies be able to develop the organizational learning necessary for scale and scope economies. Thus, managerial capitalism is the correct model in his opinion and either too much family-owner management as in Britain's case or too much involvement in financial returns as in the case of the American conglomerate movement in the 1960s and 1970s are detrimental to the long term growth of the large enterprise. The problem in Chandler's work is that of identifying the differences between success and failure cases within management dominated enterprises (as discussed above).

Mergers, acquisitions and bankruptcy: the Anglo-American model

The two other answers see the governance mechanism in financial institutions. In both the U.S. and the U.K. institutional ownership of shares has augmented individual ownership in share markets, making the share market more sensitive to professional portfolio evaluation. If a manager fails to earn the maximum return on the company's assets, the value of its shares on the stock market will be depressed. This would reduce the manager's value on the labor market for managers. In addition, it would set up both the possibility and the incentive for someone to buy the shares at depressed prices, gain control of the company, fire the manager, and through better management, increase the value of assets and company shares.[33] This would not only benefit the takeover organizer, but it would also benefit all other share owners, some of whom would have sold their shares during the takeover offer.

Indeed takeovers have been an especially important aspect of capitalism since the late 19th century. In particular, the massive merger waves in the U.S. and the U.K. during the post World War II period was marked by takeovers in the form of direct bids to the shareholders rather than as merger negotiations between managements or boards of directors. Since this was often done as a contest with and in opposition to the existing corporate direction, the process is tagged 'hostile takeover'. Hostile takeovers have also marked the restructuring and divesting of assets that has tended to reverse part of the effects of the merger movement since the late 1980s. The dramatic effect of takeovers on the industrial organization of the U.S. and the U.K. has led to a huge output of empirical and theoretical research to evaluate its effectiveness.

A recent summary of this evidence concludes that it would seem to lend scant support to the view that the takeover mechanism on the stock market selects only the efficient firms for survival or that it leads overall to a more profitable allocation of society's assets.[34] Shleifer and Vishny recently published a model that shows systematic bias in pricing of short compared to long-term assets, which would suggest that leveraged buy-outs and hostile takeovers can lead to the elimination of long-term investment projects.[35] Yet other empirical research suggest that there is no evidence showing buy-outs are associated with lower expenditures on R&D, investments or employment. Also, it has been shown that the resulting increases in debt:equity ratios are in stable industries (e.g. food and tobaccos exactly where higher ratios might be expected.[36]

Bankruptcy

The other disciplinary force comes from the possibility of bankruptcy. The function of bankruptcy is seen as a filter to eliminate economically inefficient firms and to redirect their resources into better alternative uses.

The U.S. Bankruptcy Law provides a choice of liquidation ('Chapter 7') or reorganization ('Chapter 11'). In liquidation a court-appointed trustee shuts the firm, sells its assets, and the court distributes the assets to creditors according to a predetermined priority. Under reorganization typically existing managers remain in control and the firm in operation. A reorganization plan must be adopted which settles all pre-bankruptcy creditors. Equity holders generally face large losses.

A recent analysis of incentives and practices of the U.S. law throw its effectiveness into strong doubt. Troubled firms might choose liquidation when it would be better if they continued operating, while they might continue operating even when it would be better to use their resources elsewhere. Too many failing firms are likely to remain in the same line of activity as when they were making losses. Further analysis suggests that shortcomings may be inherent in the bankruptcy procedure itself. In this case, only intervention before bankruptcy might prevent the inefficiencies noted.[37]

Bank dominated governance: the German-Japanese model

A distinct alternative to the Anglo-American system is the continental system.[38] Its most evident case is the German one, which has the following characteristics:

1   Relatively few German companies have shares listed on the stock markets. According figures cited by The Economist, slightly more than 600 or about one fourth of the AGs (*Akteingesellschaften* - public limited companies) have listed shares. At the same time the total capital of all AGs is less than the combined capital of the private limited companies (the GmbH or *Gesellschaft mit beschränkter Haftung*).

2   Unlike the American and British systems where pension funds are invested widely, in Germany common practice is for pensions to be funded in the shares of the employing companies. This significantly reduces the possibilities for outside pressures, either through participation on company boards or external divestment of shares.

3   Outside takeovers are inhibited by the common practice of placing an upper limit on the holdings of any shareholder.

4   Major resort is made to bank loans for all financing requirements and much less to either stocks or bonds. In 1970, 1980, and 1989, according to figures presented by The Economist, German non-financial companies used bank loans for 90 per cent or more of their net funds raised (by contrast, the share of bank loans for French companies dropped in the same years from 80 per cent to under 60 per cent).[39]

5   A major role in company governance is played by the universal banks, the most important of which are the big three: the Deutsche, Dresdner, and Commerzbank, and a few large regional and cooperative banks. They offer the full range of commercial and investment banking, plus brokerage services. Their importance comes not only from their role as *hausbank* in providing the most important source of financing. They also hold in the shares of their brokerage customers. Even though they are also subject to limits on their own holdings of shares, they acquire and vote the proxies of shares held for brokerage customers.

17

6   The monitoring of company behavior can be done on a relatively continuous basis by the loan officers. A question can be raised about the possibilities of doing this through members of the supervisory boards since they typically meet only twice a year. In addition, professional bankers are unlikely to intervene in technical or marketing details of the companies. One suggestion is that the banks play a special role in times of financial crisis, especially if companies must be restructured.

Even though Japan and Germany have laws with provisions similar to Chapter 11 of the U.S. Bankruptcy Act, they are seldom used. Instead, a private arrangement is engineered by the banks and it typically takes place over a rather long period of time. During this time dividends and employment might be cut, but spending on investment and R&D maintained. At the same time, in contrast to the American case where management often remains during the process of troubles, in the German situation new management is typically brought in. The whole process reduces the ultimate likelihood of failure and also reduces creditors' potential losses.[40]

According to data from the *Monopolkommission* for the late 1970s, banks held 145 seats on the supervisory boards of the 100 largest German industrial companies, or 9.8 per cent of the total (far more than the banks' own shares in these companies). The banks were represented on 61 of the companies and supplied 20 chairmen and 6 deputy chairmen of the supervisory boards. Of the 145 seats, 94 or 64 per cent belonged to the 'big three'.[41]

Privatization process and problems

The privatization process already has lasted more than three years, especially if it is dated back to the events before 1990. Since then, of course, it has acquired a new political basis and a much more serious scale. The analysis of the process, in either a positive or a normative sense, requires a difficult combination of looking at both the political processes and constraints and the technical problems and their influence on what agents involved in privatization have been able to do.

Political issues and the institutional framework

As stressed above, changes in effective property rights had certain forms and a certain momentum at the time of political changes in 1989-90. The new political forces started to redefine these processes out of several motives and with several objectives in view: to break with the past system, to impose certain equity standards, and to control the process in terms of one or more alternative definitions of what served the respective national interests. How much consideration was given to the more technical problems of privatization in these cases remains a question for research. In any case, it is rather evident that maintaining the political demand for privatization cannot be taken for granted.

Several issues have dominated the political process. Early in the transition, the possibilities of a third way or market socialism and so-called 'nomenklatura privatization' were dominant. Later the technically more complicated issues of restitution and of the role of voucher programs became important. At the present time, both the equity of privatization and the possible effects of privatization on the overall economic situation seem most important. The first two deserve some extended treatment because they led to the design of state regulated programs of privatization in all the transition countries.

Anachronism of market socialism. Discussion begins with market socialism for a special reason. Although the idea was first popular in the early non-communist opposition as well as with the communist reformers, it did not find popular acceptance.[42] Nearly every country's program soon came to accept and even stress the necessity of a private property system.[43] Yet, if we look at the actual working systems three years later which system rules de facto? Market socialism. The majority of assets are still controlled either by legal old style SOEs or commercial companies with a majority of state ownership.

Market socialism undoubtedly was first rejected out of purely political motives, if those exist. For most people, it was too close to the past and too close to the reforms they were promised, but never given. At the same time, market socialism is the de facto system not necessarily because people want it but because it has been far more difficult to privatize than expected. It may also be the case that in the face of problems of the last three years there is greater willingness to compromise with de facto socialist property than in the first breath of political opening in 1989. This remains to be seen.

Both practically and theoretically market socialism provides an interesting transition system. That is, if the idea of continuing socialist ownership could have been accepted and worked out, the transition reformers might have concentrated on creating markets and then tended to the complex problems of privatization when markets were functioning. That this has in fact actually happened is all the more important to consider some of the problems of such a system.

One set of problems has to do with the problems of market socialism in a Marxist political system. As discussed above, these would be:

1   the tendency of central authorities to intervene in specific enterprise affairs;

2   the ideological hesitation to accept profit criterion; and

3   the difficulties of entry and exit. Since the political transition, these problems have by no means disappeared, but there is more hope that they can be controlled.

Can there be a socialist capital market? How, as the agents of the state as owner of property, can the managers be given incentives to maximize the value of assets? A focus on the property question is made by Nuti, who sets out to design a socialist capital market. He assumes an institutional setting that does away with the public choice problems and some of the weakness of market forces. All planned interventions in markets by planners are to be parametric to enterprises, i.e., not enterprise specific, and all relations among state agencies undertaken on the basis of contracts.[44]

The capital market mechanism consists of the opportunity of the manager of an enterprise or an outsider, another enterprise manager, to declare a value of the enterprise assets higher than its book value. This will then result in either a revaluation or a sale of the assets to the highest outside bidder. The added value is both taxed and entered into the bonus base of the manager. The tax rate is higher than that on ordinary profit, while the bonus rate is lower than that on ordinary profit. This, it is claimed, provides the necessary incentive for managers to use capital assets efficiently.[45]

A review of Nuti's scheme in terms of contemporary contract and transactions cost theory is provided by Williamson. He finds it weak on two points.

First, the assumption of parametric intervention denies any credible behavior on the part of state authorities. They have incentives for opportunistic behavior in terms of both political gains and possible hidden material gains. Second, with less than perfect knowledge, those who bid may find themselves having bid too much. If the manager is held to account, current profits would suffer. But then the manager would have an incentive for opportunistic behavior by shifting some deferrable expenses. Alternatively, the enterprise might be pushed into some form of bankruptcy, but bankruptcy with socialist ownership only means the losses are going to be diffused to the population as a whole. In short, only gains would be taken by managers, but losses could be diffused to the public owners.[46]

Nomenklatura and insider privatization. Nomenklatura privatization was another important political issue that arose out of the way the former regimes were attempting to decentralize their economies. Strictly speaking nomenklatura is a term that refers specifically to people who worked in important positions to which they were appointed by the central apparatus of the communist party. Nomenklatura privatization refers to their efforts to acquire control and eventually income and alienation rights in their own name, or in names of family and friends, by transferring state assets first to a legal form that was free of direct supervision by a central agency and

20

eventually from there to appropriate the assets as personal or private property.

The issue has been most corrosive in Poland where the phenomena arose as a result of the loose controls over the administrative and managerial elite during the conflict in 1989 between the Party and Solidarity. Not only were the nomenklatura seen to be alienating property, they were also seen in league with foreign investors to whom sales of assets were made at low prices and nomenklatura compensated also with new jobs. Along with 'third way' ideas, nomenklatura privatization dominated the institutional reform programs of the liberals, Balcerowiez and Bielecki, and contributed to Mazowiecki's political defeat in 1990.[47]

In Hungary the process has been a lesser issue, but more common. The rather advanced organizational reforms of the former regime permitted state enterprise managers to set up associated joint-stock and limited liability companies, which, in effect, took over assets of subordinated units, factories or departments. Owners of the new units were the founding SOEs, other SOEs, banks, usually state-owned, as well as managers and top professionals of the former units. The latter had been involved in earlier independent contracting organizations internal to the SOEs and become active also in organizing authentic private and independent companies, sometimes using new cooperative forms. In some cases, these new private units acquired assets of the former state units at advantageous prices.[48]

Late 1989 and early 1990 also saw a negative political reaction to nomenklatura privatization. The Hungarian Democratic Forum used it as an election issue against the more liberal parties, but then reduced and eventually dropped the issue.[49]

The political reactions to nomenklatura privatization have been dominated by equity issues. Those who had benefited by acquiring positions of influence and high incomes under the former system, sometimes by badly compromising themselves politically and morally, were now perceived as getting rich by taking advantage of their inside information in legal, semi-legal and illegal means. The issue has carried over in most cases well past the initial transition so that the new elites are seen as nomenklatura as well and getting rich from their inside political positions.

The equity issue has clouded the efficiency issues, which are quite complex. Obviously, the effects on efficiency are negative if someone misappropriates assets that are converted into foreign bank accounts to support activities outside of the country. Also, any conversion of assets to a lesser level of productivity is inefficient, no matter who owns the assets and no matter what the transfer price was. But the mere taking of assets at low prices or at no price only redistributes wealth, it does not destroy it for the society concerned. If the transfer improves the productivity of the asset, then society gains, no matter what the effects of the transfer on the distribution of wealth.

The issue of the nomenklatura has clouded another question. This is the question of the scarcity of persons with administrative training and skills. While the nomenklatura have had defective training and experiences from the point of view of managing enterprises in market economies, they might still be the best, indeed perhaps the only persons available. In any case the issues of their comparative qualifications with non-nomenklatura is not easy, but it is only made more difficult by inserting political qualifications in a way reminiscent of the former regime.

Approaches to regulated privatization

The political responses to real or supposed nomenklatura privatization took the forms of imposing a ban on changes of organization and leasing or other forms of asset transfers of state property, creating a regulatory agency to oversee privatization, and passing legislative mandates to define desired limits to the process. As a result of all of this activity, a large menu of alternative approaches to privatization has been opened up. The following list attempts to identify the common criteria that can be used to evaluate alternative means and programs for privatization. No effort is made to list them in terms of importance.

Increasing credible commitment and political acceptability of a privatization program. An effective privatization program must be carried out. Of course, the privatization program is not the only source of political stability, but if a government cannot sustain a program or if a new government feels obliged to redo a program, that not only leads to delays that can be costly (see decapitalization below) and higher administrative costs, but also rational expectations that other programs will not be carried out.

What makes a program sustainable is still a question. Some opinion suggests that features such as voucher distribution or reduced price employee shares will enhance acceptability.[50] This would suggest that the most acceptable privatization program would try to have something in it for all the important actors.

In spite of advocacy by sympathizers with workers' management, it is hard to find evidence that provisions for free or discounted employee shares have generated positive political acceptability. Workers may understand the risk of investing their own savings and so would only be interested in highly-price discounted or free shares. The problem is that few would want shares in loss-making or in uncertain enterprises. General distributions would be highly inequitable not only because of the differences in profitability that have little to do with past worker efforts, but also because of differences in capital intensity that would mean those in labor-intensive enterprises would receive smaller ownership values than those in capital-intensive enterprises.

On the other hand, voucher schemes have not been popular everywhere, as witnessed by the continued problem of getting Poland's program underway. In the case of the voucher program in Czechoslovakia, a good case can be made that the voucher program became acceptable as a result of the persistence of the government, that is its credible commitment in the face of initial opposition and evident lack of acceptability.

Generating information. At the onset of privatization, information is very scarce so a good privatization program is one that stimulates information for those who will make decisions. Information asymmetries exist about financial conditions and technological conditions, especially for relational specific assets. Accounting and auditing is unreliable. Even to insiders information is lacking about domestic conditions and international markets.

Auctions with competitive bidding on restructuring plans as well as price generate needed information. Bids should be required for packages of complementary and relational specific assets. The optimum scope of bids should consider strong horizontal or vertical interdependencies among enterprises.[51]

Nevertheless, requiring competitive bids, especially with transparency of bidding information, might discourage strategic investors. In those cases where important information can be gained by both sides, bilateral negotiations acquire the form of a relational contract.[52]

Generally delaying privatization would increase information because of a growth in the local information services industry, accounting and consulting companies, although it increases the risk of decapitalization.

Improving the quality of post-privatization management and governance. A general distribution of vouchers is considered to promote weak post-privatization governance because it weakens the possibility of a dominant owner. In this respect, free vouchers have been seen as worse than vouchers for a fee.[53] Free or discounted employee shares are also considered a problem in this area.[54]

Voucher programs are improved when they are combined with schemes for investment funds that might acquire dominant owners' incentives for governance. A question remains about how to best organize investment funds. Should they be restricted and regulated or left free to compete? Also, should the recipients of vouchers be free to invest in all funds or directly in enterprises or should they be forced to hold shares in investment funds?

As an alternative to restrictive investment fund schemes, which have some public choice disadvantages (see below), it might be simpler to auction off one block of shares to a dominant owner along with voucher distributions. This would also reduce some disadvantages of worker shareholdings.

Reducing public choice problems. The general nature of public choice problems has already been discussed. Generally, the longer it takes to privatize, the more likely public choice problems will arise.[55] This is another reason that speed is desirable (also see decapitalization).

A system of restricted and regulated investment companies, which play a role as holding companies, can also generate public choice problems. For this reason, it has been recommended that they have built-in liquidation procedures.[56]

Probably public choice problems are enhanced when government is dependent on state-owned enterprises for revenues.[57] It makes little sense for the government to attempt to maximize revenues from privatization. And it is a common fallacy to think that the society necessarily would lose if assets are given away or privatized at a price less than true asset values. If the assets remain under domestic ownership, then privatization below real value only changes the distribution of ownership. In addition, under privatization the real productivity should rise as compared to the situation under state ownership.

Reducing administrative costs of privatization. Something quite close to the public choice problem is the behavior and cost of administering the privatization program. The agency for privatization can be seen as a substitute for capital market institutions. It is initially a public good and therefore is correctly supported by the state budget.

Initially the agency is very costly. It must invest much in learning how to do the job and, at the same time, must take the place of almost no existing capital market institutions. With the passing of time, however, the agency should become more efficient. Also, private capital market institutions increase. As this happens, more of the agency's costs should be shifted to the private benefactors of privatization and in effect create competition between it and the private capital market institutions. At some point, the work of the agency should be finished and it should be terminated or turned into a permanent capital market regulatory agency.

Initial agency costs can be reduced by privatization techniques that encourage competitive bidding and participation by private agents and managers of enterprises to be privatized.[58] Initial agency costs can also be reduced by delaying privatization while private capital market institutions, including consultants, accountants, etc., develop.

Generating new investment. Capacities for investment in transition economies are limited because of the small amounts of accumulated savings, low incomes and reduced profit opportunities. One of the virtues of transferring assets for free or at reduced prices as opposed to full price sales is that it leaves the new asset owner with greater capacities to invest. The effect on the economy, of course, depends on what is done with the proceeds of a sale. The government could make the funds available for restructuring loans, in which case the administration costs would be lost, or it could use the funds to support current expenditures, in which case all funds would be lost.

An even more effective system would be one of competitive bidding on the basis of participation shares and an offer of investment for reorganization. The participation share is the share of ownership to be taken by the investor with the rest to be taken by the state as silent partner. In the case where investors are risk neutral, this would generate more investment than a free distribution. In case of risk averse investors, it would also generate more revenue for the government than a cash sale.[59]

Avoiding decapitalization. It is sometimes said that an important criterion of privatization is speed - faster privatization is better. One of the reasons for faster privatization is to avoid the decapitalization of existing enterprises. There are two ways this can take place. As Sachs points out, a less visible, but more pervasive form is when workers and management increase wages above productivity to the detriment of cash flows needed to keep facilities or productive stocks maintained.[60]

A second form, that due to asset stripping during nomenklatura privatization, is usually mistakenly identified (as it was in the article by Sachs, quoted above). If an asset is removed from an existing enterprise and is privatized, redistribution of wealth takes place, but decapitalization from the point of view of the economy does not necessarily take place.

In such a situation, decapitalization would take place in one of two conditions:

1   assets were sold abroad and the receipts consumed; or

2   assets were removed and put into productive uses that over their expected lifetime would result in lower income flows than in their original use.

Of course, it is possible for these to happen not only with nomenklatura privatization, but with other forms of privatization.

Is there a unifying concept? An optimum privatization program might be considered one that maximized the returns on available assets over a given period of time, including transactions costs and public choice costs. Consider the expected values of the discounted net returns over, say, a period of two generations (on the grounds that people are willing to make sacrifices today for the future of themselves and their children). The correct program should seek to maximize that net value.

There are a number of trade-offs that appear to reduce losses today, but possibly increase losses in the future. Thus, rapid privatization would be expected to reduce decapitalization and public choice costs, but might risk breaking up an enterprise that would have long run viability. Other arrangements have up front costs that are high, but in return offer likely gains in the future. This might be the costs of preparing for a mass privatization with an issue of vouchers. Or the costs of a well organized widely announced auction of companies. The problem with trying to find a single approach to evaluating programs, as observed by Rausser and Simon, is the heterogeneity of SOEs and country circumstances.[61]

Among other problems, it appears that the trade-offs noted above could be quite different when considering privatization of small enterprises compared to very large enterprises. For example, it would be relatively easy to transfer small enterprises to owner-managers through leveraged buy-outs without encountering post-privatization governance problems. As a matter of fact, the greatest success of both Hungary and Poland is in this direction. By contrast, privatizing large enterprises by either country's privatization agency, even with expensive help from outside consultants, has proven very difficult.

Special problems

Problems of governance institutions in transition. We have discussed the problems of the separation of ownership and control in the traditional market economies that arises in the cases of companies with a large number of small-share owners, none of whom has an economic incentive to undertake the costs of monitoring the management. Such ownership configurations are very likely to arise with mass-privatization programs that distribute ownership widely among many recipients of coupons or vouchers. Similar configurations can result when ownership is turned over to a large number of workers.

The problem in transition countries is that initially neither the banking system (in the inspiration of the German model) nor the stock markets (in the inspiration of the Anglo-American model) are strong enough to exert governance. The banks usually are burdened with bad assets and may be technically bankrupt. Financial restructuring is going very slowly. Banks need to be, too, if they are to undertake governance functions. Although new bankruptcy laws have been passed, governments seem hesitant to use them.

Share markets and stock exchanges are weakly developed with few shares and very narrow market participation. Financial and market expertise is still needed.

Bleak outlook for the privatization of large enterprises. There are a number of reasons to expect very slow progress in privatizing the largest enterprises of CEE. It would help if the governments concerned were more willing to try to break the big enterprises into smaller pieces for privatization. Doing so is politically difficult even when it is technologically possible because employees in the weak components naturally fear being shut down if their unit is separated. Also, in some cases, in those branches characterized by Chandler as available to scope and scale economies (automobiles, heavy machinery, chemicals, computers, etc.), it might be argued that big enterprises should be kept intact.

At the same time, however, Chandler's message for these enterprises and their governments is not a very happy one. To quote him:

> The actual economies of scale or scope, as determined by throughput, are organizational. Such economies depend on knowledge, skill, experience, and team-work - on the organized human capabilities essential to exploit the potential of technological processes. *The actual economies are, therefore, the result of a long and complex learning process* [emphasis added].[62]

In writing directly about a possible remedy for Soviet enterprises, he warned that it took the Japanese nearly two decades to develop the skills and knowledge to compete in international markets, while the case of Britain suggests that it can take even longer.[63] This is the same message given by proponents of evolutionary economics.[64] If that were not enough, there is now evidence from countries outside the CEE region that simple privatization is not enough to change the behavior of former SOEs. Restructuring, particularly of management and organizational structure, may be vital as privatization is initiated.[65] That is also part of the message from evolutionary economics.

The case is even more black if consideration is given to the possibilities of using financial institutions of either of the above described models to enhance the capabilities of large enterprises. There is not the means of properly discussing why neither share and bonds markets nor banking institutions are presently in a position or are likely soon to be in one to help governance.

If faced with a future of two decades of weak management and heavy subsidies for the capital intensive industries of Chandler's case, then it would be better to either sell such enterprises to foreign investors, as is being done in the automobile industry, or break them up, sell them off and shut them down. This, after all, would leave the region still with an array of more labor intensive industries which can be more easily privatized and which still could provide growth prospects.

Remaining questions

There are clearly many more important questions raised by this review of the transition from state property to private property than have been answered. Among the issues where further exploration and research would be useful, are the following:

1 Strictly within the narrow privatization process, more research is needed to account for the relative success of some approaches such as

liquidation in Poland and decentralized processes in Hungary, while case by case privatization of large firms by the privatization agencies always meets obstacles. More research should be undertaken that compares auctions and bilateral negotiations in terms of information, cost, and efficiency of the resulting companies after privatization.

2   More comparative analysis of the legal framework is needed, especially of marginal variations in the critical laws on companies, contracts and bankruptcy and their likely effects on the efficiency of privatization, both as a process and in terms of the effective use of post-privatized assets.

3   Privatization and post-privatization operations of large former SOEs appear to pose special problems. More needs to be known about their behavior. Especially useful would be quicker circulation of costly research like that of the World Bank project covering SOEs in Poland, Hungary and former Czechoslovakia. More careful comparisons of Chandler-type enterprises in the NICs might enlarge the menu of possibly applicable lessons (this is not to be taken as any simplistic transferability of institutions or policies, as has been too often done). It would be especially important to look at possibilities for down-sizing large SOEs in capital intensive sectors and marginally shifting them into more labor intensive operations and products. Smaller enterprises seem to be easier to privatize and have fewer post-privatization governance problems.

4   The behavior of investment funds (as in former Czechoslovakia) and holding companies (as in Romania) should be explored as we gain experience, along with more research on the comparative development of private capital institutions and financial services.

5   Several different public choice problems should be further explored. What determines access to subsidies before, during and after the privatization process? How should the behavior of privatization agencies, as regulatory agencies, be explained? Do different designs of them affect the effectiveness of privatization?

6   There is need for more comparative analysis of privatization among the CEE countries. Countries have applied different approaches, some perhaps bent by particular country situations, but still extending our general knowledge. Countries are in different stages, especially with regard to legal framework, private financial services and private capital market institutions. There is still a need to better conceptualize the comparative degree that different countries are actually institutionalizing new behaviors predicated on markets and private property.

Notes

1. Typically this involves problems in the valuation of both assets and passives. State-owned enterprises typically did not have ownership as a category of passives, but did have various kinds of debts or liabilities.
2. A major difference is whether restitution involves the actual property that was originally taken or involves some form of compensation in money or kind.
3. Coase, Ronald H. (1960), 'The problem of social cost', *Journal of Law and Economics*, no. 3, pp. 1-44. Also see Cooter, Robert D. (1987), 'The Coase Theorem' in Eatwell, John, Milgate, Murray and Newman, Peter (eds), *Allocation, information and markets*, Macmillan, London, pp. 64-70.
4. When necessary, the conventional differentiation of 'the enterprise' as an ownership-organization unit and 'the establishment' as a technological unit that can be a component of the enterprise will be used.
5. Chandler, A. D. Jr. (1992), 'What is a firm?', *European Economic Review*, no. 36, pp. 483-492. The principal-agent firm is associated with Hart, O. (1989) 'An economist's perspective on the theory of the firm', *Columbia Law Review*, no. 89, pp. 1757-1774; the transactions cost firm with Williamson, O. (1985), *The economic institutions of capitalism*, Free Press, New York, and Coase, R. (1937), 'Nature of the firm', *Economica*, no. 4, and the evolutionary with Nelson, R. and Winter, S. (1982), *An evolutionary theory of economic change*, Harvard University Press, Cambridge MA.
6. The reference is to one of Coase's papers that won him the Nobel Prize in Economics in 1991; Coase, R. (1937), 'The nature of the firm', *Economica*, pp. 386-405.
7. An example is a franchise contract.
8. The alternative rational decision-making approach explains limits on the size of firms differently, as the cost of a complete sequence of short-term contracts, and represented by Milgrom, Paul and Roberts, John (1990), 'Bargaining costs, influence costs, and the organization of economic activity', in Alt, J. and Slepsie, K. (eds) (1990), *Perspectives on positive political economy*, Cambridge University Press, New York, pp. 57-89. These differences are discussed in Williamson, Oliver (1991), 'Economic institutions: spontaneous and intentional governance', *Journal of Law, Economics and Organization*, no. 7, special issue, pp. 172-183; and Wiggins, Steven N. (1991), 'The economics of the firm and contracty', *Journal of Institutional and Theoretical Economics*, no. 147, pp. 610-617.
9. Site specificity arises when work stations are located next to each other in order to reduce inventories and transportation costs, physical asset specificity is represented by transaction specific tools and dies, and human asset specificity is represented by learning by doing. Williamson, Oliver E. (1986), *Economic organization*, Harvester Wheatsheaf, New York, p. 143.
10. The essence of the problem is said also to involve differences in the timing of performance of the two parties.
11. Williamson, Oliver E. (1985), *The economic institutions of capitalism*, The Free Press, New York, p. 62, p. 260.

12. Chandler Alfred D. Jr. (1992), 'What is a firm?', *European Economic Review*, no. 36, p. 489.
13. Chandler, Alfred Jr. (1977), *The visible hand*, Belknap/Harvard University Press, Cambridge MA and Chandler, Alfred Jr. (1990), *Scale and scope*, Harvard University Press, Cambridge MA.
14. Idem, p. 489.
15. Nelson, Richard R. and Winter, Sidney G. (1982), *An evolutionary theory of economic change*, Harvard University Press, Cambridge MA.
16. Chandler Alfred D. Jr. (1992), 'What is a firm?', *European Economic Review*, no. 36, pp. 487-488.
17. His replies to discussants in a special review colloquium on *Scale and Scope* that was published in *Business History Review*, 1990, no. 64, 1990, p. 58 (henceforth referred to as 'Colloquium').
18. Colloquium, p. 750.
19. *Scale and Scope*, no. 89, pp. 285-300.
20. The criticisms of Scherer, Frederic M. in 'Colloquium', pp. 693-697.
21. Fishlow, Albert in 'Colloquium', pp. 726-729. Fishlow's remarks about developing countries also bear on the transition economies:

> The domestic managers of state enterprises and even those of private firms did not achieve their positions as a result of victories won in the marketplace. In some countries they were likely to emerge from the military; in virtually all, they would have had extensive public experience ... The theory of rent-seeking posits that the managers who would be most successful in this climate of state activism would be those who were especially privileged by public intervention.

22. Hidemasa Morikawa also says the same about Japanese *zaibatsu* and *keiretsu*: "The behavior of enterprises within the same industry in Japan is fiercely competitive." Colloquium, 722.
23. Colloquium, pp. 714-715.
24. According to one study, small firms:

1   are a source of considerable innovation activity;
2   provide considerable turbulence that makes markets more competitive;
3   promote international competitiveness through newly created market niches; and
4   generate most new employment growth.

See Acs, Zoltan J. and Audretsch, David B. (1993), *Small firms and entrepreneurship: an East-West perspective,* Cambridge University Press, Cambridge, p. 2.
25. He gives the example of the Italian government-owned conglomerate, IRI, as one of the world's largest and most diversified enterprises and also one of the most unsuccessful in terms of competitive strength.

26. Masahiko, Aoki (1990), 'Toward an economic model of the Japanese firm', *Journal of Economic Literature*, no. 28, p. 3.

27. Colloquium, p. 741.

28. Possibly by applying social learning theory.

29. The following discussion is heavily dependent on Putterman, Louis (1992), 'Ownership and the nature of the firm', *Journal of Comparative Economics*, vol. 17, no. 2, June, p. 243.

30. Laffont, Jean-Jacques and Tirole, Jean (1991), 'Privatization and Incentives', *Journal of Law, Economics, and Organization*, no. 7, pp. 84-105.

31. Laffont and Tirole also mention other kinds of firms, such as private franchise or lease operators of government property.

32. Berle, A. and Means, G. (1993), *The modern corporation and private capital*, Macmillan, New York.

33. The incentives are both indirect and direct. By providing a means of firing a manager, the manager is then exposed to competitive market forces, and a new manager can be hired whose real income is matched by productivity. If the former manager gave too generous of terms to employees or failed to earn maximum returns on cash assets, these actions could also be corrected. Better management could mean restructuring of finance, mergers, divestment of assets and subsidiaries, as well as new strategies and changes in products and processes.

34. The statistically most pronounced difference is size of enterprises, suggesting that large firms are not usually taken over, but take over smaller ones. Only rather small differences in profitability were found among the three groups.

35. Shleifer, Andrei and Vishny, Robert W. (1990), 'Equilibrium short horizons of investors and firms', *American Economic Review*, pp. 148-153.

36. *The Economist* (Survey of International Finance), (27 April 1991) p. 46.

37. White, Michelle J. (1989), 'The corporate bankruptcy decision', *The Journal of Economic Perspectives*, vol. 3, no. 2, pp. 129-151.

38. Useful institutional details for each country can be found in Hendrie, Anne (ed.) (1988), *Banking in the EEC: Structures and sources of finance*, Financial Times Business Information, London.

39. *The Economist* (Survey of International Finance), (27 April 1991), p. 46. A comparison of capital market financing in the single year 1987 by American and European non-financial corporations shows the stock (as opposed to the bond) share as 17 per cent for the American companies and 67 per cent for the European. This comparison probably reflects the special circumstances of junk bond activity in the U.S. Figures cited by Walter, Ingo and Smith, Roy C. (1990), 'European investment banking: structure, transactions flow and regulation' in Dermine, Jean (ed.), *European banking in the 1990s*, Basil Blackwell Ltd, Oxford, p. 108.

40. *The Economist* (Survey of International Finance), (27 April 1991), p. 25 and p. 49.

41. Cable, John (1985), 'Capital market information and industrial performance: the role of West German banks', *The Economic Journal*, p. 119.

42. As Kornai points out, some form of market socialism became the objective of reformers in Hungary and other countries, including Poland (Kornai (1990), *The road to a free economy*, Norton, New York, p. 57.) Among those favorably

disposed to ideas of employee ownership in Hungary, were the Christian Democratic and Social Democratic parties (Bartlett, David, (1992), 'Political economy of privatization: property reform and democracy in Hungary', *East European Politics and Societies*, vol. 6, no. 1, p. 106). In Poland, according to Winiecki, workers management was the favored economic institution of the Solidarity group that negotiated the Round Table agreement that eventually led to the first non-communist government. In spite of continuing strength of the idea among many in the Solidarity camp, the government's program in October 1989 pushed for privatization. In Winiecki's opinion, it was not until the law on privatization was passed in the spring of 1990 that Poland clearly rejected a 'third way' solution (Winiecki, Jan (1992), 'The political economy of privatization' in Siebert, Horst (ed.) (1992), *Privatization: Symposium in Honor of Herbert Giersch*, Tubingen, Mohr, pp. 76-78. Also, Stark, David (1992), 'Path dependence and privatization strategies in East Central Europe', *East European Politics and Societies*, vol. 6, no. 1, p. 39.

43. Indeed, Kornai goes further: "... the basic idea of market socialism simply fizzled out. Yugoslavia, Hungary, China, the Soviet Union and Poland bear witness to its fiasco." *The road to a free economy, p. 58.*

44. Nuti, D. M. (1989), 'Feasible financial innovation under market socialism', *Journal of Comparative Economics*, pp. 94-95.

45. Nuti also provides two other institutions that could be phased in: a share market for state agencies, but not individuals, and an options market to allow individuals to benefit from knowledge of enterprise performance. Ibid., p. 95 and pp. 98-102.

46. Williamson, Oliver E. (1991), 'Economic institutions: spontaneous and intentional governance', *The Journal of Law, Economics and Organization*, vol. 7, Special edition, pp. 159-187.

47. Stark, David (1992), 'Path Dependence and Privatization Strategies in East Central Europe', *East European Politics and Societies*, vol. 6, winter, p. 17.

48. Op. cit., pp. 44-45; also, Bartlett, pp. 97-104.

49. Op. cit., p. 92.

50. Bartlett, p. 86.

51. Maskin, Eric S., (1992), 'Auctions and privatization' in Siebert, Horst (ed.), *Privatization: Symposium in Honor of Herbert Giersch*, Tubingen, Mohr, pp. 115-141.

52. Stark, p. 43.

53. Stark, p. 33.

54. Kornai says that schemes of free distribution are mistaken, but it is not very clear why. He might be saying that it reduces interest in how assets are managed (*The road to a free economy*, pp. 81-82).

55. Stiglitz makes the point that privatization increases the transactions costs of obtaining government subsidies and so helps the government's commitment to competition and hard budget constraints. Stiglitz, Joseph E. (1991), 'Theoretical aspects of the privatization: applications to Eastern Europe', *Working Paper*, Institute for Policy Reform, Washington DC, p. 15.

56. Fischer, Stanley (1992), 'Privatization in East European transformation' in Clague, Christopher and Rausser, Gordon C. (eds), *The Emergence of Market Economies in Eastern Europe*, Blackwell, Oxford., p. 241.

57. The problem arises for two reasons. Such sources of revenue were built in the inherited system and it has proven difficult to develop new tax systems. Also, tax exemptions for new private enterprises plus their ability to avoid reporting taxable activities makes it difficult to rely on direct business taxes or even sales or turnover taxes.

58. How much privatization should cost is an unanswered question, as far as I know. Poles were generally very unhappy at the fact that private consultants had charged an average of 25 per cent of the privatization proceeds of Poland's first case by case privatizations, but nothing has been said about whether this was too much, normally expected or a bargain.

59. Demougin, Dominique and Sinn, Hans-Werner (1992), 'Privatization, risk-taking and the communist firm', *Discussion Paper Series*, no. 743. CEPR, London.

60. Sachs, Jeffrey D. (1992), 'Privatization in Russia: some lessons from Eastern Europe', *American Economic Review*, vol. 82, no. 2, pp. 43-44.

61. Rausser, Gordon C. and Simon, Leo K. (1992), 'The political economy of transition in Eastern Europe' in Clague, C. and Rausser, G., *The emergence of market economies in Eastern Europe*, Blackwell, Oxford, p. 352.

62. Colloquium , 737.

63. Chandler, Alfred D. Jr. (1992), 'Organizational Capabilities and Industrial Restructuring: A Historical Analysis' *Working Paper*, p. 36.

64. See Murrell, Peter (1992), 'Evolution in Economics' in Clague, C. and Rausser, G., *The emergence of market economies in Eastern Europe*, Blackwell, Oxford, pp. 35-53.

65. Krueger, Anne (1992), 'Institutions for the New Private Sector", in Clague, C. and Rausser, G., *The emergence of market economies in Eastern Europe*, Blackwell, Oxford, pp. 219-223. Krueger cites the case of Turkey in particular, one which might be useful to compare in the CEE region.

2 Privatization, company management and performance: a comparative study of privatization methods in the Czech Republic, Hungary, Poland and Slovakia

Valentijn Bilsen[1]

Introduction

In the transition from a centrally planned economy to a market economy, the privatization of the existing state-owned enterprises (SOEs) was one of the first policy measures that was taken. It can be considered as a necessary condition to reach the final goal of a market economy. But it is not a sufficient condition. Privatization in the former centrally planned economies is necessary because it is virtually impossible to develop a market economy only on the base of newly set up private firms that must compete with a vast amount of incumbent state firms that produced at the eve of transition almost 100 per cent of total output and allocated and controlled vast amounts of resources. Furthermore, SOEs operate under different conditions than private firms. Soft budget constraints, their monopolistic 'market' position and their preferential connections with the state banks are only some of the factors that would make it almost impossible for a new private firm to enter the market[2]. The role of privatization is to reallocate the production of goods and services to private owners who are all subject to the same market rules, and to introduce an appropriate governance structure that makes the former SOEs respond to market signals.

The privatization of SOEs alone will not guarantee the development of a market economy. The scale of goods and services that is needed by firms and families to operate adequately in a market economy is much wider than the scope of output that the existing SOEs and government services produced. One of the most striking examples is the development of the capital market. For example the stock market, different types of debt financing and credits, leasing, insurance, bankruptcy law, and credit screening are institutions which were previously almost nonexistent and which are necessary inputs for a private firm.

The aim of this chapter is to investigate the relation between privatization, company management and the performance of the enterprise. Privatization and company performance influence each other. First, privatization determines future company performance. It generates a new ownership structure of the companies and in the economy as a whole. The new ownership structure creates a new set of incentives for the management, which in turn leads to a change in company performance. Second, the expected company performance determines the outcome of the privatization process as well. To analyze this relation, the concept of adverse selection is introduced. Given the state of capital market development, it helps to explain the difference in outcomes of different privatization methods that are used. As a consequence it also sheds light on the perceived difference in privatization paths of several countries. From an evolutionary point of view, the result of privatization generates the starting conditions of the new-born market economies. The outcome of privatization gives more insight into the future evolution of these economies.

An overview is made of the privatization methods in the Czech Republic, Hungary, Poland and Slovakia. These are the countries that started first with privatization. The emphasis is laid on ownership structure and management issues. The overview is presented in the second section. Before embarking on this overview, a theoretical analysis is presented in the first section on corporate governance in a transition economy. The last analyzes the effect of expected performance, as measured by profitability, on the outcome of privatization. Therefore, the concept of adverse selection is introduced. The precondition for executing the investigation is a definition of privatization.

In this study privatization is defined as the activity that aims to transfer ownership rights from the state into private hands. The ownership rights include the right of income, control and alienation of the property. As a consequence, privatization is considered an instrument of organizational reform and of marketization. The creation of new enterprises is not considered a form of privatization,[3] but as another way of creating a market economy, which is by no means less important than privatization. But from an analytical point of view it is useful to distinguish the two concepts rather than to generalize.

36

Privatization and company performance in a transition environment

Privatization and company performance are interrelated with each other in multiple ways. The interrelation can be explained as follows. On the one hand, privatization determines the future company performance. On the other hand, the expected company performance influences the privatization outcome. Various channels can be detected through which privatization influences company performance and vice versa. In the following section, the first channel will be discussed briefly in the context of an economy under transition. In section six, adverse selection and privatization, the relation between expected company performance and privatization is discussed.

Ownership control

The influence of privatization on company performance can be abstracted as running through mainly two channels. The first channel is ownership control on the management, and as a consequence on firm performance. The change of enterprise ownership from the state to private individuals leads to a change in the set of incentives for the management. This causes a shift in management behavior, which in turn has its implications on company performance. One can distinguish several incentive sets according to the mutual relation and identification of the owner and the manager.

The basic case is that where the manager fully owns the firm. In this case, there is no principle-agent problem between the owner and manager since both are one and the same person. The governance of the firm is done within the organization of the firm. The manager's income and wealth depends directly on the firm's profit and performance. As a consequence, one might expect that the maximization of the profit of the firm is the major target of the manager. He will try to allocate company resources to achieve this goal.

When the manager and the owner are not identical, several possibilities arise. In the case of privatization the owner can be a private person or a private legal identity. The private person can be an employee or a personal investor. The private legal identity can be another firm or an institutional investor. Often the government retains a part of the assets as a means of control. This is the well known textbook case of corporate governance. The owner of the firm wants a maximum value for his shares. This is not necessarily the prime concern of the manager, since his income does not directly depend on the value of the shares. He might be more interested in a maximum market share, growth of the firm, diversification, fringe benefits and of course his managerial salary. However, the manager who is involved with the daily command of the firm often possesses more information about the firm than the owner does. The asymmetry in information together with the difference in goals creates a principle-agent problem between the owner and the manager. The shareholder has to bear extra costs to get the information that the manager has and which is needed to evaluate whether the management actions are in his interest. For small shareholders, the governance cost might be too high in comparison to the benefits from control. They hope that another shareholder bears the cost and as a consequence act as a free rider. This illustrates the importance of a core investor (or a 'real' investor) for whom it is beneficial to bear the governance costs.

In established market economies, the governance of management behavior is mainly performed through asset markets and credit markets[4]. The typical feature of economies in transition is that these markets still have to be developed and are in their infancy. As a consequence capital market control does not yet fulfill the monitoring function as in a full grown market economy.

In employee-owned firms, the principle-agent problem between owners and managers is solved as well. The employees both control and manage the firm. Ben-Ner, states: "... the scope for cooperation is larger and that for conflict (is) smaller than in capitalist firms."[5] The employees will be more keen to contribute to the performance of the firm when they receive together with their wage a part of the firm's profit.

As long as managerial decisions do not include the hiring and firing of workers, this might be true. The employee not only invests his human capital or labor into the firm, but also his savings. The distribution of his wealth is less diversified and he faces a higher risk. On the one hand, this might generate a greater incentive to undertake actions that support the firm's performance. On the other hand, the mobility of the workers is reduced. This makes firing an employee a more severe decision than in other forms of ownership, because the whole wealth position of the employee deteriorates. Compared to other forms of ownership, this leads to sub-optimal decisions. Furthermore, if the fired employee keeps his shares in the firm, he becomes an outside owner and the possibility of conflict rises.

A crucial condition for enterprise governance in employee-owned firms is the homogeneity of the group of employees. One might expect that over time the income and wealth position of the employees is diverted. Especially when employment in the particular firm they own is not the only source of income. As the homogeneity of the group of employees decreases, different views on company control and management will emerge. This reduces the scope for cooperation and might lead to sub-optimal company performance.

Competition

The second channel is the control on management via the output market in which the firm is operating. The degree of competition on the output market determines the set of actions for the manager and as a consequence also the desired performance of the workers. On the input side of the firm

38

the development of the capital market and the labor market, both for the labor force and managers, determine the performance and conduct of the firm. The market does not only provide limitations to certain strategies, it also provides opportunities that stimulate the firm's growth when they are met with the right set of actions.

A central actor in the transition is the government. Its task is to introduce the appropriate legislation which is needed to operate a market economy in a generally accepted way. For instance: the commercial code, company legislation, bankruptcy law, and intellectual property law are only some of the necessary rules. Another task is to remove the factors that were inherited from the centrally planned economy and which hamper the operation and development of a market economy. Price decontrol, trade liberalization, and demonopolization are only some of the various tasks that have to be done to allow a market environment to develop.[6] The quality of legislation and the speed at which it is issued determine the potential for market development. Beside the role of the government as a regulator, it also acts as an owner of SOEs that must be privatized.

Overview of privatization methods with respect to their ownership structure and management in Poland

Preparations for privatization

Commercialization. Commercialization is the conversion of a state-owned enterprise into a joint-stock company or into a private limited company. The shares are owned by the State Treasury. All the rights and duties of the former SOE are assumed by the new company. After commercialization, the shares will be sold to private individuals or entities by means of another privatization method. As such commercialization can be considered as a preparatory step before the change of ownership.

The typical triangular structure of the former SOE, also known as the 'Bermuda Triangle', is replaced by a corporate structure.[7] The new corporate structure consists of a supervisory board and a board of directors. The supervisory board represents the sole shareholder: the Treasury. The board of directors is appointed by the Treasury and is responsible to the supervisory board. The goal is to introduce a clear management structure that increases the firm's viability in a competitive market environment.

Commercialization is used by the government as an instrument to prevent the SOE management from selling the firm's assets. The board of directors is authorized to stop the sale of assets. Asset stripping may occur for various reasons. Bienkowski (1992), states that asset stripping is used by the SOE management as a tactical measure to prolong the existence of the SOE. The assets are sold as a survival measure to meet salary and credit payment deadlines. By selling the assets, the SOE management gains time.

It can preserve its beneficial status, which will be lost after commercialization. Asset stripping diminishes the value of the firm and makes it less suitable for privatization.

Liquidation on the basis of Article 37 of the State Enterprise Privatization Act of 1990. The state enterprise is closed and part or all of its assets and liabilities are sold or leased to a new company. This new company can be established by outside investors, by the management or by the employees. The new company might also be a joint venture with a domestic or foreign partner. A combination of these methods is also possible. Liquidation under Article 37 can be considered as a preparation and a means to enable privatization. The initiative to apply this method is taken after the joint decision of the state enterprise's management and workers. Also the Ministry of Privatization and the local authorities can initiate the process. This method of privatization is mostly used for small and medium-sized enterprises. The average employment in firms that were privatized under this method is 360.[8] Simoneti (1993) states that in most cases the assets are leased. The annual rent amounts to a minimum of 20 per cent of the asset's value. At the end of the leasing period, negotiations on the acquisition of these assets are started. Research done by the Ministry of Privatization revealed that in five per cent of the firms that were privatized by this method, the management fears bankruptcy because of the rents paid for the lease of the assets.[9] This was particularly the case with firms privatized in the beginning of 1990. The valuation of the assets at that time was too high compared with the present financial situation.

Liquidation on the basis of Article 19 of the State Enterprise Act. This method most frequently occurs when the enterprise cannot pay the obligatory dividend to the Treasury as a consequence of long-standing losses. So, the main reason for applying this kind of liquidation is insolvency. Most possibilities occur for the transformation of assets as with the previous liquidation method, except the lease of assets. Liquidation on the basis of Article 19 is also called forced liquidation, while liquidation on the basis of Article 37 is labelled as voluntary liquidation. In both cases the state enterprise is crossed off the register of SOEs.

Privatization methods

Small privatization. Local municipalities and communes sell or lease small businesses to the public. The businesses are retail shops, restaurants, trading firms, construction firms and transport enterprises. The local authorities decide on their own the specific terms of sale and lease contracts, e.g. how the valuation is done, whether to hold an open auction, can employees acquire property at preference terms? The decentralization produced a wide variety of procedures, but on the other hand led to great progress. Most privatized shops are leased or rented because of the lack of capital.

Initial public offering (IPO). After commercialization, the shares of the new firm are sold on the stock exchange. As a consequence the property of the firm can be widely distributed among the public. This often creates the problem of having a strategic investor. An IPO can only be applied to joint-stock companies. The appointed companies must have good financial results and a well known positive image. Only a limited number of enterprises meet these conditions. The success of an IPO depends heavily on a well established stock market. At the start of the program the Warsaw Stock Exchange was not yet established and the sales of shares were done at bank offices. The procedures for application at bank offices were time consuming. They caused a slow down of the IPO execution. Another factor that hampered the process was the complexity of the procedures and techniques. The outcome was also determined by the poor private savings.

Direct sale of enterprises. This method is also referred to as 'initial private offering', 'trade sales' and 'sales by negotiation'. The shares of limited liability companies and joint-stock companies established by commercialization are sold to one or more strategic investors. The privatization ministry appoints an advisor who gathers information about the company. He invites domestic and foreign potential bidders by direct contact and through advertisements. Potential investors must state the price they are prepared to pay. Additionally, they must submit a business plan in which the investment and employment policies are specified. The selected investor pays only a part of the actual value to the State Treasury. The remainder is financed with a long-term credit, backed by the active assets which will be paid by future profits. Only a limited number of firms were privatized with direct sales. Foreign investors supplied an important part of the capital. The direct sale of enterprises was the main route for foreign investors to acquire former SOEs.

Mass privatization. Almost two years after the idea of 'mass privatization' was launched, the Polish parliament adopted the Act on National Investment Funds and their Privatization on 30 April 1993. Several substantial amendments were included. Below, the adopted version of the mass privatization program is explained.

Each adult Polish citizen can acquire shares at a certain fee, from one or more of some 20 newly created National Investment Funds (NIFs). These investment funds have participations in the enterprises that are selected to be privatized by the mass privatization method. They are managed by experienced domestic and foreign firms. The fund management firms will be selected after an international open tender. An NIF is established as a joint-stock company, that is, controlled by the general meeting of shareholders, an executive board and a supervisory board. The executive board's main task is to increase the value of the NIF by making investment decisions on behalf of it, and by controlling management of the companies in which they have an important share. The supervisory board will be responsible for appointing members of the executive board and will sign the management contract with a foreign or Polish managing firm, which it will choose. The supervisory board's president must be a Polish citizen, as must be two thirds of its members. In case the NIF is fully owned by the Treasury, all members of the supervisory board must be Polish citizens.

The enterprises eligible for mass privatization are medium-sized, financially sound SOEs which have already been commercialized. The Minister of Privatization proposes a list of enterprises that can be included in the mass privatization program. The enterprises get 45 days to express their decision to participate or not. If they do not respond within this period, they will be automatically selected for mass privatization.[10]

Sixty per cent of the selected firm's shares will be allocated to National Investment Funds. Up to 15 per cent will be distributed for free to the companies' employees, 25 per cent remains in the hands of the State Treasury, and 55 per cent of the shares allocated to the NIFs (which equals 33 per cent of total shares) will be held by one single NIF. It acts as a leading shareholder. The remaining 45 per cent (or 27 per cent of the firm's total assets) are equally distributed among the other NIFs. They act as minority shareholders in that particular firm. The shares of the privatized firms can be sold on the stock exchange. So after trading, an NIF can change its stake in the firm's assets and its control on company management. Foreign investors are allowed to acquire shares in the NIFs in accordance with Warsaw Stock Exchange regulations.

The Minister of Privatization gives each adult citizen an opportunity to obtain a share certificate that represents in its term a share in each NIF. The registration fee for the certificate costs maximum ten per cent of average monthly wages. The certificate (also called MPP share or investment certificate) is tradable throughout the country. It can be traded on the Warsaw Stock Exchange. It is also possible to exchange one MPP share against one NIF share. The NIF shares are tradable on the Warsaw Stock Exchange as well.

The mass privatization program still has to be carried out. On 18 March 1993, the Parliament rejected the Government's program for mass privatization. In this program 600 SOEs were included. On 30 April 1993, the Parliament adopted a second version of the mass privatization program. The Senate ratified the program on 7 May 1993. In the new program the 600 SOEs are privatized in two phases. In the first phase 200 SOEs will be privatized. This concerns mainly enterprises which are already commercialized and in full hands of the Treasury. The investment certificates are distributed for free to the former employees and to retired persons to compensate for their loss of purchasing power. Five to eight NIFs will be established to acquire shares in those companies. The second phase consists of the privatization of 400 SOEs. In this phase, about 15 NIFs will be established.

After the elections, the new government created a selection committee consisting of 13 members who must select the companies that offered to manage the NIFs. This will be an enormous task, since 7,000 offers were received. Out of these 200 members must be selected who will manage the NIFs and who will sit in their controlling board.[11] It might be expected that this will create a bottleneck in the execution of the MPP. It might also be interpreted as evidence that the government wants to carry out the MPP in a rather centralized way.

The weak point in this program is that for effective control the main investor still needs the cooperation of the Treasury which holds an important amount of the shares. The success of mass privatization with respect to company performance will also depend on the willingness of the fund managers to act as real investors. If they behave as a passive institutional investor who is only interested in the dividend of the shares, the value and profit opportunities of the firms will decrease. As a consequence the NIFs will undergo the same fate, since their portfolio value will decrease. The final victims will be the many persons who invested in the NIF certificates. On the contrary, the strategic investor controls the firm's management and takes a lead in determining the strategic goals of the firm and supports this by new capital when needed.

The result of mass privatization will also depend on the trading of the enterprises' shares between NIFs. Trading allows an NIF to acquire a higher stake in the firm and to improve its control. This might lead to a less diversified NIF portfolio, generating higher risk. To compensate for the increased risk, the NIF expects a higher return from the firm under control. Therefore it will monitor more carefully the firm's management and conduct.

Sectoral privatization. Strictly speaking, this is not a distinct method of privatization. It is a method of privatizing a group of enterprises at the same time. Several enterprises can be offered as one single package to investors. As of December 1991, the Ministry of Privatization selected 143 enterprises grouped in 34 manufacturing sectors to be privatized.[12] For each group an advisor, mostly a large international consulting firm, is chosen, who makes an analysis of the whole sector as well as of the individual firms. He devises in cooperation with the firms a privatization and a restructuring strategy for each individual enterprise. He makes propositions to ensure their viability and to attract foreign capital.

It is expected that sectoral privatization would improve the choice of the privatization method and the way of restructuring. Also the implications for market concentration are more readily perceived. Weaker businesses can be aggregated with stronger ones in a single package. Although sectoral privatization looks nice on the drawing table, practical execution was not a success.

Reprivatization. The legislation on the restitution of assets taken over by communist authorities still has to be approved by the Parliament. The government proposes a combination of restitution in kind, where possible, and equity bonds as substitute property in the other case. Regional courts would determine the right to restitution and the form of it. The absence of a legal solution for restitution retards the execution of other privatization methods. As an explanation for the slow progress of reprivatization, Mizsei points out that before transition, the land was predominantly in private hands. Therefore its restitution was not a political target.[13]

Overview of privatization methods with respect to their ownership structure and management in the Czech Republic and Slovakia

Preparations for privatization: commercialization

The SOE is converted into a joint-stock company whose shares are owned by the National Property Fund. Unlike in Poland and Hungary, commercialization is not always a necessary step before privatizing an enterprise.

Privatization methods

Small privatization. The assets of small enterprises are sold by public auction. It is also possible to sell directly to persons who already leased the enterprise. Only private entrepreneurs and companies can participate in the auctions. The state enterprises are not allowed to take part. Foreigners can take part only in cases when the auction is repeated. This happens when the property was not sold in a first round of bidding. Any person interested in a certain property is allowed to propose what should be auctioned to the regional privatization commissions. These commissions are regional bodies of the Ministry of Privatization (MoP). It is the MoP that makes the final decision about the property eligible for selling. The regional commissions organize the auctions.

Simoneti reports that in 75 per cent of the cases the assets were leased on a long term basis, while in the remaining cases the assets were sold.[14] The newly privatized businesses cannot be sold for the next five years. Normally, the debts of the old enterprise are not transferred to the buyer. They are paid off out of the sales proceeds. Small-scale privatization is mostly applied in services, trade, buildings and small factories. Note that agriculture and health care are privatized within the large-scale privatization scheme.

Large privatization. This form of privatization can be done using several techniques. These include:

1   public auctions;

2   public tenders;

3   the founding of joint-stock companies;

4   the direct sale to a selected buyer;

5   free of charge transfer.

Large-scale privatization is organized by the MoP. Anyone, including foreigners, can submit a proposal for the privatization of state property to the local or state authorities. This is called a 'privatization project'. In turn the authorities submit their standpoint to the MoP. The MoP does its own analysis and chooses the best proposal.

The 'privatization project' must contain information about the method of privatization, the estimated value of the firm, and the future legal form. Also the method of division of shares for a joint-stock company, the method and form of determining the price in case of a sale, and a time schedule has to be specified.

In the beginning of large-scale privatization the standard used for selection was the business plan and the investment commitment. Later the price offered and the techniques of auctions and tenders were preferred. The sale and transfer of the approved projects is performed by the special privatization agency of the National Property Fund. Large-scale privatization is applied to industrial firms, construction companies, agriculture, transport, banking telecommunications, health care, culture and education. The participation of employees is limited to five per cent of the shares in large firms and to ten per cent in smaller ones.

Voucher scheme. The state offered to citizens over age 17 an equal amount of investment vouchers. To take part in the voucher scheme, the citizens had to register and pay a registration fee. The vouchers could be used to bid for shares of firms that entered into the Voucher Privatization Scheme. The vouchers could also be placed with one of the 437 investment funds. The investment funds manage a portfolio of shares of privatized enterprises. When the property is sold or when no more vouchers are available, the whole process stops. This process is called a 'wave'. The next wave starts with a new registration of the participants and with a new list of enterprises to be privatized by this method. After the end of the voucher privatization, the participants will receive a certificate confirming that they are the rightful owners of the firms the shares of which they ordered.

The investment funds have the legal form of a joint-stock company. Each fund cannot hold more than 20 per cent of its shares in any one company. The portfolio of each fund must contain at least ten businesses. Government officials cannot be board members.

The first wave of voucher privatization was finished at the end of 1992. Both republics have sold shares in 1,491 companies. In the Czech Republic only 6.8 per cent of the shares remained unsold. In Slovakia this percentage was 8.2 per cent.[15]

The degree of public participation in the voucher scheme was very high. Altogether 8.54 million individuals took part in this privatization process. This amounts to 80 per cent of those eligible to take part in the scheme. Seventy-two per cent of the points were invested in investment funds. Only 28 per cent was directly invested. Nine voucher investment funds ended up controlling about 50 per cent of all investable points. The remaining investment funds controlled about 25 per cent of the points. Six of those major investment funds are controlled by banks, and one by an insurance company. As a consequence, ownership is highly concentrated and it is mainly controlled by the banking sector. Company governance may approach the German model where the banks act as investment banks and the loan officers closely monitor company behavior.

Restitution. Real estate that was nationalized or confiscated between 1948 and 1989 was returned to the former owners, but involved only buildings and land. Securities, money and goods were not subject to restitution. Movable assets (e.g. works of art) were returned only when they still existed. If the property could not be given to the former owner, he could receive financial compensation either in cash or in the form of shares.

The period for submitting a claim was limited. For agricultural land it lasted from June 1991 till 31 December 1992. For other assets it was only six months beginning in June 1991. Property transfer is done directly between the former and the new owner.

Overview of privatization methods with respect to their ownership structure and management in Hungary

Preparations for privatization: commercialization

An SOE can be converted into several corporate forms before privatization takes place. The intention to prepare an SOE for privatization must be reported to the State Property Agency (SPA). The business plan, the objectives and the valuation of the assets have to be specified. Within 30 days the SPA approves, amends or rejects the proposal. In the enterprises that are transformed, the SPA is the owner and controls the management.

Privatization methods

Small privatization, retail privatization, or pre-privatization. In the fall of 1990, Parliament passed a law on privatizing retail firms, restaurants and small service firms which have less than 15 employees. Some of the enterprises have already been leased before by private persons. The firms are sold in open auctions to the highest bidder. Foreign buyers are excluded. The SPA organizes the auctions, and determines the minimum price. Potential buyers are the former managers and employees, and outside investors. The new owner is not allowed to liquidate the business for a set period after the purchase. The majority of purchases were financed by special low cost loans with a maximum repayment period of ten years.

Self-privatization or decentralized privatization. The privatization process is initiated by the enterprises themselves or by an investor. The forms of privatization are:

1   transformation of state-owned firms into joint-stock companies;

2   associations with private partners, both domestic and foreign;

3   sales.

The execution of the privatization is decentralized, and is done by consultant firms that are selected by the SPA after a tender. The consultant firm invites bids for the company to be privatized and selects the best offer. They receive five per cent of the proceeds of the sale after the investor has paid for the privatized firm. If the process proceeds quickly, an additional premium of up to five per cent is given. After the completion of the sale, the SPA does an evaluation. The quality of documentation, the pace of execution, the price of the sale and the number of transformed companies are some of the elements that are considered.

This form of privatization primarily involves small and medium-sized companies in processing industries. It usually occurs when a strategic foreign investor appears.

There are two stages (or waves) of self-privatization. In the first stage, lasting from September 1991 until 31 March 1993, it was planned to privatize 421 firms with the aid of 84 consulting companies.[16] The maximum size of the firms was 300 employees and HUF 300 million annual sales. The second stage started in August 1992 and lasted to the end of 1993. It was aimed at larger companies with 300-1,000 employees and HUF 300 million-1 billion sales revenues. In the second wave, 272 companies were to be privatized and 129 consulting firms were involved.[17]

This privatization method is considered as one of the most successful in Hungary. By 31 January 1993, 57 per cent of the companies listed in the first wave were transformed. For 19 per cent of the firms a sale was started or completed,[18] and 16 per cent of the enterprises listed in the second privatization wave were transformed by 31 January 1993. Sale was started for two per cent.[19]

Large privatization. To privatize large enterprises, several methods are used. The first privatizations were 'spontaneous' without much control of the state. The enterprise councils of the SOEs were allowed to form mixed or privately-owned companies. The valuation and the sale were not subject to control or oversight by the state. In many cases this led to asset stripping. The most profitable parts of the firm were sold privately, leaving the state with an empty shell. Managers who feared for their job sold (parts of) the enterprise well below its earnings value on the condition that they retained their position. In too many cases the old nomenclature seemed to benefit from this practice. To stop the abuses, a better supervision and control of the privatization process was needed, which led to the creation of the SPA. The SPA proposed several new methods to avoid previous problems and to speed up the privatization of large enterprises.

1　Management-initiated privatization, or controlled spontaneous privatization. The managers of the firm initiate the privatization and take an active role in the process. The SPA imposes a standard procedure and does a close follow-up. The SPA must give its final consent before any sale of assets or shares can take place. The valuation of the assets is done by a consultant firm that is approved by the SPA.

2　Investor-initiated privatization. Outsiders can bid for companies over the head of existing managers. The bid is done directly to the SPA without any negotiations with the target company. The investors also have to specify a business plan for the enterprise. In all cases the SPA encourages other investors to make rival bids. The SPA decides autonomously. This form of privatization can lead to a hostile takeover. The SPA checks the view of the management and the founder. If the management is opposed, the Agency can take over the control of the firm and finish the privatization. Because of the asymmetry in information between the firm's management and the SPA, a hostile takeover seems to be unlikely.

3　Active privatization or state-initiated privatization. The privatization process is initiated by the SPA. The agency publicly announces the list of the state companies put up for sale. Then, consultant firms, banks and accounting firms make a bid to act as an advisor for the privatization of the selected companies. Foreign consultant firms are allowed. In fact most of the bids come from American and West-European auditing and consulting firms. The SPA selects the advisors. The consultants accomplish the valuation of the firm and propose an appropriate privatization method. They also make an assessment of the business management, the market position and the technical situation of the firm. The process from announcement to final privatization is called a 'privatization program'.

Up to now there are already three programs. In the first privatization program the goal was to sell 20 firms before the end of 1992. It concerned large well-known firms with a better business result than the average firm. By the end of 1991, ten companies were transformed and the sale of four companies was started.[20] The planned SPA share in each company is on average 36 per cent.[21]

The second program consisted of two phases. In the first phase, 12 large state-owned shell companies were to be sold. During the period of uncontrolled spontaneous privatization, these firms sold more than 50 per cent of their assets into one or more joint ventures. While the shell firms themselves were not viable, the joint ventures are. In the second phase 11 more companies were added, but no progress in privatization has been made.

The third program consists of group privatization programs for several sectors. The program included 31 construction firms and 15 companies in agriculture. However, almost no progress was made in this program.

Employee share ownership program (ESOP). The ESOP is a privatization technique that is designed to encourage the participation of employees in the firm's capital. Employees can buy assets or shares of their company financed with loans. The purchase is not guaranteed by their personal property, but by that of the enterprise. The employee becomes the owner to the extent that he repays both principal and interest. If the employee is not able to repay the loan, the assets are transferred to the bank or to the government. The conditions for applying an ESOP are: first, the existing owners agree with the program, second, at least 50 per cent of the employees assume responsibility for it, and third, a financial institution is prepared to grant a loan for the purchase of the shares.[22] The employees have to pay 25 per cent of the purchased shares in cash.

Compensation of confiscated properties. Unlike in other countries (Poland, former Czechoslovakia), former owners whose properties were confiscated by the communists have no right to claim back their properties. Instead, financial compensation will be given in the form of freely tradable, interest-bearing compensation bonds.[23] The Compensation Law came into effect on 11 August 1991. The bonds can be used to buy privatized state property. Only the original holders of the compensation bonds can use their certificates to buy farmland and apartments. They can also convert their certificates into life annuities. Holders who bought the bonds on the secondary market can only buy assets and shares in privatized firms sold off by the SPA.

Adverse selection and privatization

Adverse selection and the profitability of the firm

One of the basic issues in all privatization methods is the uncertainty about the future environment and the presence of asymmetric information about the expected performance of the firm. The main criterion for company performance is profit. Profit is the basis for self-financing, it determines the future existence of the firm and the welfare of those whose incomes, goods and services are produced by it. After serious simplification, for the sake of analysis, one can distinguish three agents that are present in each privatization method. The first agent is the government who owns the firm to be privatized. The second agent is the manager who ran the enterprise before privatization. The third agent is the potential outside investor. Let us assume that the agents interact in a one-shot game.

The government wants to sell the firm at the best conditions. This might be the highest price. But also the employment opportunities, investment plans and future profitability are taken into consideration. Both the manager and the outside investor are potential buyers. They want to buy at the lowest price to maximize their future rents.

Let us assume that the uncertainty about the future environment of the firm is the same for all agents. However, the asymmetry of information concerns the internal conditions of the enterprise. The manager has a more precise view on the profitability of the firm than the outside investor and the government. The outside investor and the government can only obtain the same kind of information at high costs, for instance by using the services of consultancy firms, or by doing their own investigations.

According to the adverse selection phenomenon, the agent with the most accurate information, the manager, will try to get the profitable firm at the lowest price possible. If the enterprise is not profitable, it is left for the outside investor. However, the outside investor faces more uncertainty about the value of the firm so that his maximum price is lower than that of the manager. If the manager does not buy the firm it is interpreted as an indication of bad profitability by the outside investor, and he will not buy the firm.

Speed of privatization

When the aspect of time is brought in, the adverse selection theory gives a possible explanation for the fact that the firms with good performance are privatized earlier than the poorly performing firms, because managers will be the first agents that buy the most prosperous enterprises. This explains why in Hungary management-initiated privatization of large firms came earlier and was more successful than the state-initiated privatization methods. Also the initial uncontrolled spontaneous privatization of large enterprises can be considered as supporting evidence. In Poland the success of privatization by liquidation on the basis of Article 37 fits this theory as well. Part or all of the SOEs' assets were sold or leased to the agent with inside information. Considering the triangular structure of Polish state-owned enterprises, one can identify the workers' council and the trade union council as agents with inside information. During the period of January 1990-November 1992, 71 per cent of the SOEs were privatized by liquidation on the base of Article 37.[24] It was estimated that 80 per cent of the assets of these firms were leased out to employee-owned companies.

Secondly, it helps to explain why it is difficult to sell firms that started late with their privatization. According to the adverse selection theory, the most profitable firms are privatized first by the managers. The outside investor knows this as well. When the former SOE management takes a long time to initiate the privatization process, or when the SOE is forced by the government to start privatization, the outside investor will interpret this as a sign of low profitability. He will be less keen to buy the enterprise.

Sectoral privatization and self-privatization

Third, adverse selection gives a rationale for sectoral privatization programs on the condition that each group has the same distribution of profitability of its firms. Suppose an investor wants to acquire a strategic position in a certain branch of industry. Within each branch or group, packages are offered that consist of assets from several enterprises. The package can also be a combination of enterprises. If each package has the same distribution of profitability, the uncertainty of the outside investor is reduced. Since each package contains both profitable and less profitable enterprises and assets, the outside investor is prepared to pay a higher price. The privatization process is speeded up. This is the case in Poland. However, Hungarian sectoral privatization does not seem to fit this scheme. This might be explained by the small number of large enterprises involved in each group or sector. The smaller the number, the less diversified the portfolio of the investor is, and the lower the price he is willing to pay.

A second aspect that supports the sectoral privatization programs is the availability of information. In the Polish case, the sectoral studies of the consultant firms are available to all agents involved. This reduces the information bias. For the outside investor the uncertainty of buying a less profitable firm at a price that is too high is reduced. Also the government has a better idea about the price it can ask for its property. This increases the likelihood of reaching a 'market' price that is acceptable for both the government and the buyers, which are the managers or the outside investors. Ceteris paribus, privatization will proceed faster.

Evidence for this argument is also found in the success of the Hungarian self-privatization program. The introduction of consultants improves the information symmetry. This allowed foreign strategic investors to participate more easily in the privatization process, which in turn led to an amplification of it. Furthermore, this not only leads to an injection of capital, but also to the introduction of new management techniques and know-how. It is an additional boost to the privatization process. The domestic investor can also benefit from this situation. He receives the same information from the consultant as the government, the manager and the foreign investor. Additionally, competition between the manager, the foreign outside investor and the local outside investor leads to a search for the optimal allocation of resources. One can imagine that this improves the quality of privatization.

Small privatization

In all countries considered, small privatization is one of the most successful privatization methods. Jones reports that up to the summer of 1992, in Poland more than 80 per cent of the retail shops and road transport services were privatized.[25] The figures for construction, wholesale trade and import traders were respectively 50 per cent, 60 per cent and 40 per cent. In the Czech Republic and Slovakia small privatization was completed for 74 to 84 per cent, depending on the estimates of the total number of units eligible for this kind of privatization.[26] In Hungary, by 31 December 1992, 74 per cent of the units lined up for pre-privatization were privatized.[27]

One of the reasons for the success of small privatization might be that the problem of adverse selection is present to a much lesser extent than in the case of large privatization. Adverse selection generates the result that the most profitable units are sold or leased first to agents who have inside information. The remaining firms are left unsold, as was explained above. However, the likelihood that this occurs with small firms is lower than with large firms. This is because the cost of sampling information of a small firm is much lower than for a large enterprise, because the internal structure is not as complex as that of a big one. It is less costly to analyze.

Lower information costs create less asymmetry between the agents. The outside investor faces less uncertainty and he can make a better assessment of the firm's profitability. As a consequence, he is more willing to participate in the privatization process. And a market for small firms will emerge.

Second, for a small unit it is often easier to find alternative applications. This reduces the probability of an unfortunate purchase. The alternative costs of a small enterprise are lower. The outside investors are more keen to invest. This reduces the effects of the asymmetric information.

Conclusions

Privatization, company management and performance

1   Mainly in Poland and Hungary, the beginning of privatization was marked by asset stripping. SOE managers, who had inside information, bought the best parts of the firm, leaving a less viable company in the hands of the government.

2   The success of initial public offerings was limited, because of the lack of savings and established secondary markets. As a consequence, the distribution of assets among the population by this typically 'western' technique was disappointing.

3   The introduction of investment funds considerably boosted the privatization process. The clearest example is the Czech and Slovak voucher scheme. Considering the quality of privatization, the main issue is whether the investment funds will act as a strategic investor or not. In the Czech Republic and Slovakia it was found that the banking sector controls a large part of the assets. The quality of the voucher privatization program will depend to a large extent on the behavior of the banks. In the Polish mass privatization program, the question is raised whether the share of an NIF in a firm is sufficient to act as a real investor? The result will clearly depend on the attitude of the Treasury and on the strategies of the other NIFs.

4   In all the countries considered, most of the small and medium-sized enterprises (SMEs) are leased on a long-term basis. This contributed to the progress of privatization. However, the government still remains the owner of the assets and has its say in the final allocation. Hence, it is premature to conclude that all SMEs are fully in private hands.

5   In the Czech Republic, Hungary and Slovakia, the restitution of property is on good course. Hungary uses the technique of compensation bonds, while in the Czech and Slovak Republics restitution in kind is also possible. In Poland, reprivatization is still an unsolved matter. This hampers the progress of other privatization methods.

6   Employee ownership is most widespread in Poland and Hungary. Especially in Poland the influence of trade unions on management is important. In the absence of a well established secondary capital market and the occurrence of high unemployment, employee-owned firms might not be the optimal solution for ownership transformation.

1   Given the low state of capital market development, the existence of asymmetric information between the manager and the outside investor causes an adverse selection problem in privatization. The most profitable firms are acquired first by the manager who owns inside information. Hence, adverse selection explains the success of management-initiated privatization.

2   Adverse selection also implies that the less profitable firms are left for the outside investors. Outside investors are not willing to buy the remaining enterprises, because they face a great uncertainty about the firm's viability. As a consequence, no market will develop for these firms. This helps to explain the slow process of many large privatization programs.

3   The introduction of consultant firms reduces the adverse selection problem, because the information is made available to all agents involved in the privatization deal. This reduces the information asymmetry and facilitates the generation of an equilibrium price.

4   Offering packages of shares with the same distribution of profitability is a method to overcome the adverse selection problem. This gives a rationale for sectoral privatization.

5   The adverse selection problem is less severe in small privatization. It is easier to find an equilibrium price for the firm, which explains the relatively great success of the small privatization programs.

Notes

1. I would like to thank Istvan Abel, Wouter Biesbrouck, Irina Dumitriu, Barbara Fakin, András Giday, Miroslav Hrncír, Marvin Jackson, Jan Klacek, Józef Kozinski, Jan Mujzel, Ognian Panov, Agnes Sári-Simkó, and Stavros Thomadakis for helpful comments. The financial support from the Belgian National Science Foundation (NFWO) is greatly acknowledged as well.
2. The coexistence of private and state firms in a transition economy is discussed in greater length by Thomadakis in this volume.
3. However, the broad concept of privatization includes the creation of new enterprises. Filipowicz (1991), describes this concept of privatization as: "... a general and gradual process consisting of two elements: the transfer of property rights to the existing productive resources from the state to private individuals and stimulating the development of new, privately-owned enterprises that could in future create the core of the emerging market economy." Mizsei (1992), defines privatization as: "... all forms of expansion of the private sector in the national economy".

4. See also Jackson (1994) and Thomadakis (1994), in this volume.
5. Ben-Ner (1992), p. 14.
6. For a more elaborate discussion and analysis of the government's role in marketization see Biesbrouck and Jackson (1994 forthcoming).
7. The triangular structure consists of: the founding body, e.g. a ministry or a local authority, the workers' council and the trade union council.
8. s.n. Polish News Bulletin of the British and American Embassies, no. 99/92, 22 December 1992, p. 7.
9. idem, p. 7.
10. Paczynski, L. S., (1993), 'Mass privatisation programme', East European Business Law, no. 93-V, p. 9-10.1
11. Mojkowski, J. (1994), 'Selekcja z klucza', Polityka, no. 6, p. 4.
12. Jermakowicz, E., Jermakowicz, W. (1993), p. 36.
13. Mizsei (1992), p. 294.
14. Simoneti (1993), p. 92.
15. The numerical data on the first voucher wave come from: s.n. PlanEcon Report, 25 January 1993, p. 1-3.
16. s.n. Privinfo, no. 5, 1993, p. 3.
17. idem, p. 6.
18. idem, p. 4.
19. idem, p. 7.
20. s.n. Privinfo, no. 4, 1992 p. 2-3.
21. idem.
22. Hollauer (1993), p. 7.
23. Mihályi (1992), p. 27.
24. s.n. Polish News Bulletin of the British and American Embassies, no. 13/93, 1993, p. 6.
25. Jones (1992), p. 115.
26. Mládek (1993), p. 3-4.
27. s.n. Privinfo, 1 February 1993, p. 24.

References

Ben-Ner, A. (1992), 'Organizational reform in Central and Eastern Europe: a comparative perspective', Working Paper, no. 8, Leuven Institute for Central and East European Studies, Leuven.
Bienkowski, W. (1992), 'The Bermuda Triangle: why self-governed firms work for their own destruction', Journal of Comparative Economics, vol. 16, pp. 750-762.
Biesbrouck, W. and Jackson, M. (1994), 'Markets and business competition: the institutional and economic framework' in Jackson, Marvin and Biesbrouck, Wouter (eds), LICOS Studies on the transitions in Central and Eastern Europe, vol. 3: Marketization, demonopolization, and the development of internationally competitive enterprises, Avebury, Aldershot, Kent.

Filipowicz, L. (1991), *Small and medium-size business development policy for a successful transition: the case of Poland*, EUR/EC Workshop on economics of restructuring, March.

Hollauer, T. (1993), 'Experience with ESOP, with special regard to the articles of association', *Privinfo*, vol. 2, no. 3, pp. 7-9.

Jackson, M. (1994), 'Property rights, company organization and governance in the transition', this volume.

Jermakowicz, E. and Jermakowicz, W. (1993), 'Approaching business valuation in the Polish Privatisation Programme' in s.n. (1993), *Valuation and privatisation*, OECD, CCEET, Paris, pp. 25-50.

Jones, C. (1992), 'Privatisation in East Europe and the former Soviet Union', *Financial Times Business Information*, London.

Mihályi, P. (1992), 'Property rights and privatization. The three-agent model: a case study on Hungary', *Eastern European Economics*, vol. 31, no. 2, pp. 5-64.

Mizsei, K. (1992), 'Privatization in Eastern Europe: a comparative study of Poland and Hungary', *Soviet Studies*, vol. 44, no. 2, pp. 283-296.

Mládek, J. (1993), 'Small privatization in the Czech Republic: the Hey-Day is over', *Privatization Newsletter of Czechoslovakia*, no. 12, pp. 8.

Simoneti, M. (1993), 'A comparative review of privatisation strategies in four former socialist countries', *Europe-Asia Studies*, vol. 45, no. 1, pp. 79-102.

s.n. (1993), 'Statistics on the first and second phases of self-privatization at the beginning of 1993', *Privinfo*, vol. 2, no. 5, pp. 2-7.

s.n. 'Poland's socio-economic situation in 1992', *Rzeczpospolita Statistics Supplement*, no. 1.; translated in: Polish News Bulletin of the British and American Embassies (1993), no. 13, 1993, pp. 2-13.

s.n. 'Privatization facts and figures', *Privinfo*, vol. 2, no. 3, 1993, pp. 24-25.

s.n. 'Results of Czechoslovak voucher privatization: world's largest 1991 IPO successfully completed: part I: overall review and statistical data', *PlanEcon Report*, vol. 8, no. 50, 51, 52, 1993, pp. 81.

s.n. 'A review of privatisation methods', *Zycie Gospodarcze*, no. 51/52, translated in: Polish News Bulletin of the British and American Embassies (1992), no. 99/92, 1992, pp. 6-9.

s.n. 'Small enterprise development through privatization', *Privinfo*, vol. 1, no. 4, 1992, pp. 2-6.

Thomadakis, S. (1994), 'Perspectives on financial systems and enterprise efficiency in the transition economies of Central and Southeastern Europe', this volume.

3 Financial intermediation and company management: the case of the Czech Republic

Miroslav Hrncír[1]

Introduction

Irrespective of differentiated initial conditions and strategies adopted, the fall of economic activity in emerging market economies (EMEs) of Central and Eastern Europe has been everywhere deeper and more protracted than expected. Identifying the causal factors of their common characteristics, both 'insiders' and foreign observers pointed to the impact of continued defects and rigidities in factor markets' functions. This applies, in particular, to financial markets and financial intermediation. The conclusions advanced in the literature underlined their constraints, suggesting that a considerable misallocation of available savings was taking place in reality. "Savings that are urgently needed to lay foundations for future growth are being wasted by banks, lending to loss-making enterprises" (Dittus, 1993, p. 15). As it turned out, a considerable share of credits extended in the course of transition proved to be bad loans, enlarging thus the volume of bad loans inherited from the past.

The argument was put forward that, due to a number of reasons, the flow of savings in the EMEs was biased in favor of large, mostly loss-making state firms, at the expense of prospective clients especially in the emerging private sector (Begg and Portes, 1992, Phelps et al., 1993). It follows that the emerging sectoral allocation of credit was claimed to support the assertion of the continued misallocation of savings, i.e. of the flows of available funds to the wrong clients and projects.

Without a doubt, the defects of financial intermediation constrained the 'real' economy's developments in the EMEs. The impact of widely spread debt-contract failures and of bad, non-performing loans was transmitted into the economy through a number of links. The misallocation of savings affected the pace of restructuring in the enterprise sphere directly, cutting down the funds and means available for prospective companies and new entrants. At the same time, bad loans and debt contract failures also worked through their impact on the formation of macroeconomic variables. The evolving interest rates and the margins between banks' borrowing and lending rates accommodated for the increasing share of bad loans and debt-contract failures. As a result, the interest rate burden was higher than otherwise could be the case, also affecting good and prospective enterprises.

The experienced drawbacks in financial intermediation were claimed to be co-responsible for the slow pace of restructuring and for the sluggish recovery achieved as yet in the EMEs. "The failure of credit markets in the Central and East European countries is the primary reason for the stalled transition" (Abel and Bonin, 1993, p. 6).

The causal links also run the other way around: financial markets, their functions and efficiency are heavily influenced by the real economy's developments. The importance of this feedback was demonstrated in a spectacular way in the course of the recession of the early 1990s. The stability and soundness of even well-established financial and banking systems in a number of developed market economies were severely undermined which resulted in banking crises in several cases, including in particular Nordic countries (BIS, 1993). The description and evaluation of the Finnish case, i.e. one of those particularly hit, is given in Nyberg and Vihriäla (1993).

Just the links and interactions between financial markets and the 'real' economy's developments in the EMEs, and not so much the state of financial markets per se, became a matter of concern and discussion. If persistent, the identified defects and constraints of financial intermediation would represent a serious comparative disadvantage of the EMEs. In such a case the prospects of their recovery and of sustainable growth would remain rather gloomy. Consequently, both external and domestic observers seemed to agree that financial markets and financial intermediation were the spheres where a further radical advance was most urgently needed. The 'second' financial and banking reform thus appeared to be on the agenda.

The discussion in this chapter is concentrated in particular on banks and on their functional efficiency and stability. This reflects the emerging consensus that the banks are and should be the core of financial systems in the course of transition, and especially in its early stages. The relevant issue is therefore what types of banks are needed and what to change in their conditions, environment and incentives to secure a more satisfactory standard of financial intermediation in the EMEs. Should the established 'old' banks continue to play their dominant role, providing their restructuring, privatization and changed incentives (Dittus, 1993) or should the newly emerging banks be increasingly relied upon and their developments supported (Phelps et al., 1993) or is it better to 'import' banking institutions from the west (Schmieding and Buch, 1992)?

A related strategic issue is what type of banking system is to be aimed at in the EMEs and what role the banking institutions could and should play in the 'real' economy and in corporate governance in the transition period. The universal type of bank, implying integration of commercial and investment banking, is increasingly given preference in the literature discussing transition economies (Saunders and Walter, 1991). This model is consistent with the trend in the EU countries as reflected in the second EU Banking Directive which provided for a broad range of banking activities. The legislation in the EMEs also seemed to mostly follow the option of universal banking, as documented for example by the Polish and Czech banking laws.

Nevertheless, there are also significant risks and constraints implied in the adoption of universal banking in transition economies. Along with the arguments of general relevance, in the specific conditions of the EMEs the given model could also imply the risk of keeping non-viable firms afloat. The re-emergence of such a traditional behavior pattern would be just the opposite of what should be aimed at in the transition countries (Hare, 1991).

Against this background, this chapter examines the properties of financial intermediation and its impact on the developments in the corporate sphere in the case of the Czech economy. It is organized as follows: the standard provided by mature market economies for financial markets and for financial intermediation in the EMEs is discussed in section two. Section three evaluates the structural constraints in financial intermediation inherited from the past. The evolving trends are examined in section four, concentrating on the role of credit and deposit markets, on the causes and consequences of bad loans and on the private sector's share in credit allocation. The chapter concludes with a discussion of the potential role of banks in the 'real' economy and in corporate governance.

What type of financial system for EMEs

The standard for the emerging financial markets and banking institutions in transition economies is provided by the developed market economies. However, the types of financial institutions and their role in the non-financial corporate sector diverged substantially both in the individual development stages and across individual countries as well. The governance functions of financial institutions are discussed by Jackson (1994).

Following Saunders and Walter (1991) three alternative modes of financial contracting and flows can be distinguished:

1   credit institutions, i.e. savings/commercial banking and other traditional forms of intermediated finance;

2   the public securities markets, i.e. investment banking and securitized intermediation;

3   private placement, i.e. various financial direct-connect mechanisms between borrowers and lenders.

These types of linkages were dominant characteristics of individual development stages in the world economy. While after World War II traditional bank intermediation prevailed, since then the financial systems of developed market economies have tended to be more market-oriented. This shift was promoted in particular by the process of financial markets' deregulation, accompanied by the widely spreading replacement of non-negotiable debts and loans by negotiable securities, i.e. by securitization. The trend towards 'disintermediation' has been further intensified by the increasing weight of privately-issued securities and related products.

Reflecting these degrees of maturity, the relevant standard for the evolving financial systems in transition economies should be the post-war stage of developed market economies, and not their current systems. However speedy the process of catching up in the institutional spheres might be, the successive development stages cannot be avoided. It may therefore be counterproductive to try to imitate, without an adequate learning and adjustment process, the present highly sophisticated financial markets and products of developed market economies.

Though some trends and initial expectations show otherwise, which prompted Corbett and Mayer (1991) to address them as "progress with the wrong model", there has been a fairly general consensus that the banks are the institutions which are in place and which must form the core of the financial systems in the EMEs.

Legacy of past financial intermediation

Structural constraints in commercial banking

Banks in former Czechoslovakia were restructured into a two tier system at the beginning of 1990. Unlike in Hungary and Poland,[2] the move only coincided with the start of radical political and economic reforms. All of the seven banks operating in January 1990 were undercapitalized and, at the same time, burdened with the inherited non-performing loans and/or with credits extended at artificially low interest rates. Assessed on the basis of international standards, all were technically insolvent.

The two largest banks, Commercial Bank (Prague) and General Credit Bank (Bratislava), which took over the established network of the former mono-bank, started their activities at a rather unfavorable capital adequacy ratio (the ratio of own capital to assets) of only about 1.5 per cent. The respective ratios of savings banks were even lower, around one per cent.

These figures highlighted the fragility of the emerging banking system. The two above-mentioned banks constituted the backbone of the entire commercial banking system; their combined share amounted to 68 per cent of total bank credits and 25 per cent of total deposits at the end of 1990.

However, the real situation was even gloomier, as there was hardly any reliable loan-risk classification applied at that time. According to the available estimates, after the start of the stabilization and liberalization program in early 1991, about a quarter of the outstanding loans could be qualified as either bad, non-performing or high risk loans.

The non-performing and risky credits were not confined to domestic clients. A specific feature of ex-Czechoslovakia was a relatively high share of frozen assets abroad, especially in the former Soviet Union and in some Middle East countries. Though nominally in better shape than other commercial banks, the balance-sheets of the Czechoslovak Trade Bank (a joint-stock company which, in the past, almost exclusively effected foreign exchange payments and deals) were also heavily burdened with questionable claims. These latter were mostly linked to various forms of bilateral trading and to government credits extended in the past.

Hence, the initial characteristics of the emerging banking sector in former Czechoslovakia were not particularly sound. The capital/asset ratio of the existing banks with mostly inherited portfolios from the mono-bank institutions was not only far from meeting the international standards, but also compared unfavorably with that of their Polish and Hungarian counterparts.[3] The banks' undercapitalization, coupled with uncertainty about the real performance of a substantial share of their portfolios inevitably implied severe constraints on the functional capabilities and behavior of existing commercial banks, in particular with respect to the defaulting clients.

Inter-enterprise arrears

The deficiencies in the banking sphere in the post-communist countries have been interrelated with the increasing burden of inter-enterprise arrears. Irrespective of the concrete conditions and policies of individual countries, their nominal volume surged in the first 'stabilization stage' of transition.

The interpretation of these developments was not, however, unambiguous. Some features were evidently similar to those of trade credits which are quite common in market economies. Referring to its dimensions in developed market economies, the argument was raised that, at least to some extent, the proliferation of inter-enterprise indebtedness in the EMEs was a spontaneous correction in response to the previous conditions under which trade credits were not officially allowed (Begg and Portes, 1992).

However, the type of monitoring and the criteria used were not capable of distinguishing 'voluntary' and 'involuntary' forms of indebtedness or of drawing a dividing line between them. The inter-enterprise arrears mushroomed within the given time-span under the pressure of deep recession, of the restrictive stance of macroeconomic policies and of the diminishing sales in both domestic and foreign markets. Consequently, they were apparently mostly of an induced character. As such, they differed from the institution of trade credit, interfering with the rules of the game and with the market-type code of behavior.

Table 3.1
Inter-enterprise arrears in former Czechoslovakia
(in billion CSK, end period)

	Inter-enterprise indebtedness			Number of firms	
	Total	Primary	Secondary	Total	Indebted
June 1989	16.3	-	-	-	-
December 1989	6.6	3.2	3.4	1,522	460
June 1990	14.5	7.8	6.7	4,728	727
September 1990	24.9	12.4	12.5	5,566	1,106
December 1990	47.8	20.0	27.8	4,827	1,651
March 1991	76.7	27.1	49.6	7,388	2,766
June 1991	124.3	35.0	89.3	8,201	3,587
September 1991	147.2	36.0	111.2	9,109	3,960
December 1991	170.7	37.8	133.0	11,043	4,723
March 1992	170.5	35.9	134.5	n.a.	n.a.
June 1992	165.2	37.8	127.4	n.a.	n.a.
September 1992	154.6	35.3	119.3	n.a.	n.a.

Source: Financial indicators, Federal Ministry of Finance.[4]

In former Czechoslovakia, the volume of inter-enterprise arrears has been increasing with some ups and downs since the early 1970s. Substantial increases were registered in 1981-82 and again in 1987, i.e. in periods when the monetary authorities tried to tighten credit policies.[5] An unprecedented sharp increase has been registered since the second half of 1990. As follows from table 3.1, the above-mentioned trend even accelerated in the course of 1991 and at the beginning of 1992. At the end of 1991, the identified sum of inter-enterprise indebtedness amounted to CSK 170 billion, i.e. to a quarter of total banks' credits extended to enterprises and 4,723 firms (from the total of 11,043, i.e. 43 per cent), showed payments arrears to other firms.

The data in table 3.1 include the split into primary and secondary indebtedness. Primary arrears or primary indebtedness is identified as the state in which an enterprise's overdue payables are greater than its overdue receivables. Secondary arrears or secondary indebtedness includes all its overdue payables which could be covered by its overdue receivables.[6] As follows from table 3.1, since the beginning of 1991, i.e. since the start of macroeconomic stabilization and liberalization programs, the increases of inter-enterprise arrears were due to the secondary form, i.e. to increasing clients' failures to honor their obligations to suppliers at maturity. Though unprecedented, the identified trend of inter-enterprise arrears was much less dramatic in real terms, when corrected for inflation (compare table 3.2).

Table 3.2
Nominal and real changes in inter-enterprise arrears (in billion CSK)

Year	Producer Price Index[a]	Increase (+) or decrease (-) of inter-enterprise arrears	
		Nominal[b]	Real[b]
1990	117.1	+41.2	+35.2
1991	154.8	+122.9	+79.4
1992	108.4	-15.1[c]	-13.9[c]

a   as against previous year
b   as against December of previous year
c   September 1992
Source: Federal Statistical Office (1992), *Bulletin*, Prague and own calculations.

As mentioned, there are no consistent official data on inter-enterprise arrears covering the recent period. This is in particular due to the extensive institutional and ownership changes taking place which precluded the continuation of the former time series data. At the same time, the discussion shifted to the overall indebtedness of economic agents, including besides its inter-enterprise component also tax and bank credits' arrears.

The estimates of such a wider coverage of indebtedness in former Czechoslovakia identified its volume around CSK 300 billion in 1992, while for the Czech Republic alone in the range of CK 170-220 billion (compare Ptácek, 1993).

With respect to the inter-enterprise arrears only, the available evidence seemed to suggest (with all the reservations about the respective data) that, since the second half of 1992, their volume in the Czech Republic not only ceased to increase even in nominal terms, but also showed some tendency to contract. Nevertheless, the extent of the problem coupled, in particular, with the bad loans in banks' portfolios continued to be a potentially serious destabilizing factor.

What made the situation in the EMEs specific

Bad loans, debt contract failures, interfirm arrears, bankruptcies and banking crises are certainly not confined to the EMEs. Both market and transition economies have recently been exposed to a number of exogenous and endogenous constraints, including the world as well as the domestic recession, external shocks and failed regulatory policies. Despite these parallels, the nature, intensity and impact of debt contract failures in EMEs substantially differed from those in developed market economies, the main distinguishing causal factor being the legacy of the past. This legacy included not only the inherited bad loans and accumulated inter-enterprise arrears, but also the institutional and behavioral legacy.

In the established regime of a standard market economy any firm has both positive and negative incentives to comply with the code of behavior requiring, among others, to honor debt contracts at their maturity. Under the rules of the game, the options are in principle either adjustment or bankruptcy and exit. Such behavior is supported by the established institutional and legal infrastructure (Thomadakis, 1994).

Unlike in market economies, these institutional and behavioral preconditions for maintaining and enforcing the market code of behavior were mostly missing in the initial stage of transition in the EMEs. The increasing occurrence of debt contract failures and inter-enterprise arrears in the EMEs were not matched by actions such as bankruptcy procedures against defaulting clients. In the given environment, neither banks nor suppliers were ready to initiate such procedures. Their apparent passivity in enforcing credit contracts (Begg and Portes, 1992) derived from a number of interrelated factors.

Because of the legacy of central planning, there was hardly any correlation of the state of the accounts of individual firms with their 'real' financial standing, and even less with their medium-term prospects. The existing non-transparency in the performance of an individual firm made it almost impossible to draw any clear-cut distinction between its illiquidity and insolvency.

Given the extent of disarrays in payments flows and of overlapping claims and liabilities from the past, the suppliers could not properly assess the payments morale and credibility of their clients ex-ante. However, even more important proved to be the inertia of traditional behavior patterns of state-owned enterprises. They were accustomed to effecting deliveries without devoting proper care to the risk of failed payments. This tendency 'to go on delivering the supplies' appeared exacerbated under the pressure of the shrinking domestic and foreign markets early in the 1990s. For example, many Czech firms were reported as routinely continuing their exports to the former USSR during 1991, irrespective of the fact that the institutional framework of the Comecon trading system was already phased out.

The debtors' conditions and responses were also controversial. On the one hand, many of them were constrained by the payments arrears of their own clients. At the same time they did not have the compelling incentives to do their utmost to meet the obligations of the contracts. Some evidence has been accumulated that, in the given environment, even the firms which were neither insolvent nor illiquid deliberately joined the 'club' of overdue debtors, which also included private firms and joint ventures. By not paying at maturity they in fact drew cheap credit, cheaper than at commercial banks.

The passivity of commercial banks in enforcing the credit contracts was in the initial transition stage also conditioned by their own fragile situation, given sub-standard capital/asset ratios, the substantial share of non-performing loans in their portfolios, along with initially nonexistent or only rather limited loan-loss provisions. As a result, the 'old' commercial banks appeared to be in a locked-in position, dependent on the survival of their major clients.

The desirable institutional and legal framework, though differently advanced across individual EMEs, could be developed and made effective only gradually. In former Czechoslovakia a step-by-step approach was also adopted in the implementation of bankruptcy legislation. The bankruptcy law approved by the Parliament in autumn 1991 was to enter into force on 1 October 1991. The extent of its enforcement was, however, constrained by a 'blocking' article, effectively allowing bankruptcy procedures to be started only in the special case of overindebtedness. Though scheduled for full applicability since autumn 1992, it was once more postponed until April 1993 and, at the same time, its provisions amended by the new reading in the Czech Parliament at the beginning of 1993. These amendments introduced a number of safeguards, including e.g. granting the institution a three months' grace period, aimed at avoiding a domino effect and an uncontrolled chain of bankruptcies.

The reasoning behind such a cautious approach was to reconcile the intensity of the given law's impact with the underlying institutional and economic conditions of the particular development stage. In the given period, the constraining factors were seen in particular in the delayed implementation of the first wave of voucher privatization and in the still underdeveloped infrastructure, including courts. That is why the amendments to the bankruptcy law provided for out-of-court solutions. The intention followed was to give way to bankruptcies, but at the same time to avoid unnecessarily high costs, given the existing institutional constraints.

The aim of both the government and the Central Bank was to maintain a balanced course and the links of their policies with the advance in privatization and in microeconomic and institutional adjustment. As it turned out, unlike macroeconomic stabilization and liberalization, to make institutional and ownership changes effective required a longer time period, and the policies had to become more evolutionary and gradualistic.

To summarize, the mass occurrence of debt-contract failures in the initial development stage of the EMEs was linked to the lacking microeconomic and institutional prerequisites, which could secure both the incentives to adhere to the conditions of the contracts as well as the means to enforce them. In the given environment, the old bad loans and debt contract failures inherited from the past interacted with the new ones, arising in the course of transition itself.

This absence of microeconomic and institutional prerequisites also constitutes the reason why bad loans and debt contract failures in the EMEs, unlike in market economies, became a widespread phenomenon which acquired a persistent and structural character, not only a cyclical one. As a result, the 'disease' proved to be long-term and contagious.

Properties and development of financial intermediation

Credit and deposit markets

As in other EMEs, the introduction of a two-tier framework marked the start of a dramatic increase in bank and also non-bank financial intermediaries. From only seven commercial and savings banks operating in former Czechoslovakia in January 1990, numbers have steadily grown, amounting to 57 licensed banks in the Czech Republic alone at the end of 1993. About half of them were banks with foreign capital, either partly or wholly foreign-owned, including subsidiaries and branches of foreign banks (compare table 3.3).

Table 3.3
Structure of the Czech banking system
(licensed banks as of 31 December 1993 in %)

Banks with domestic private participation	33.3
State financial institutions	1.8
Banks with dominant state participation	7.0
Building savings institutions*	7.0
Branches of foreign banks	17.5
Banks, foreign-owned	15.8
Banks with foreign participation	17.5

*   with foreign participation
Source: *Financial Statistical Informations*, CNB.

This rapidly increasing number of banking institutions found, however, only moderate and gradual expression in the relative market shares and, in particular, in the intensity of competition. Nevertheless, compared to the initial stage of reform in 1990, the rate of change proved to be significant. The combined share of six large (old) banks in total credits extended by the banking sector in former Czechoslovakia was as high as 98 per cent and their share of deposits was even higher, more than 99 per cent at the end of 1990 (compare table 3.4).

A relatively sharp drop to a 77 per cent share in total credits only a year later was mostly due to the transfer of a substantial part of perpetual credits for inventories from the balance sheets of large commercial banks to the newly-established Consolidation Bank. Then, however, the process of changes in the market structure accelerated. Table 3.5 identifies this tendency in the Czech credit and deposit markets in 1993.

Compared with the almost negligible activities at the end of 1990, the newly-founded small banks managed to increase their share to 24 per cent in total bank credits and to 20 per cent in total deposits. Though the degree of concentration particularly in deposit markets remained relatively high and the share of newly-established banks, including foreign-owned ones, has been increasing only gradually, the tendency to a more equal distribution and a more competitive setting in both credit and deposit markets has been clearly developing since 1990.

Table 3.4
Structure of credit and deposit markets in ex-Czechoslovakia
(1990-92, in %)

	31 Dec.1990		31 Dec. 1991		31 July 1992	
	C	*D*	*C*	*D*	*C*	*D*
Total large 'old' banks	97.9	99.3	77.1	93.3	77.0	91.2
Commercial Bank Prague	47.8	17.5	26.4	17.8	24.7	18.4
General Credit Bank Bratislava	20.1	7.9	14.5	8.0	14.7	6.5
Investment Bank Prague*	14.6	8.9	14.8	7.0	10.5	5.1
Investment Bank Bratislava	-	-	-	-	5.0	2.3
Czechoslovak Trade Bank	5.1	2.7	7.9	4.1	7.0	4.4
Czech Saving Bank	6.7	42.3	9.2	39.0	10.2	37.8
Slovak Saving Bank	3.6	20.0	4.3	17.4	4.9	16.7
Total small 'new' banks	2.1	0.7	7.2	3.5	10.6	6.6
Agrobank	1.6	0.4	2.9	0.9	3.1	1.6
Other small banks	0.5	0.3	4.3	2.6	7.5	5.0
Consolidation Bank	-	-	15.7	3.2	12.4	2.2

C   credits
D   deposits
*   in 1992 the Investment Bank was split into Investment Bank Prague and Investment Bank Bratislava
Source: Czech National Bank, own calculations.

The evaluations of the identified developments were, however, controversial. On the one hand the banking system maintained its bipolar structure, with parallel existence of large and small banks, while medium-size banks continued to be almost entirely missing. This division coincided with the split between 'old' and 'new' banks, i.e. between banks active already at the start of transition in January 1990 and those founded only in the course of transition. The 'big four' of Czech banking has been formed by the Commercial Bank, the Czech Saving Bank, the Czechoslovak Trade Bank and the Investment Bank (which was transformed through a merger into the Investment and Postal Bank in December 1993), i.e. entirely by old banks. If the five per cent share in credit and deposit markets is taken as a threshold, then only one of the new banks, Agro Bank, would qualify as one of the big banks, which implies that the Czech banking system continued to be dominated by old banks. Accordingly, it was argued, the changes in the market shares in favor of new banks were too slow.

The other side of the coin was, however, unfavorable side-effects of the growing market share of new banks. The share of 'new' bad loans in the portfolios of this group of banks was increasing while the level of reserves and loan-loss provisions did not adjust enough to match the increased risk involved. Constrained by the evolving limits, the credit expansion of small banks appeared to slow down in the first months of 1994.

In a parallel to market structure, the maturity pattern of bank credits also continued to be biased compared to what could be considered both rational and desirable. Data in table 3.6 show the persistently growing share of short and medium term credits in the past years at the expense of long-term ones. This shift resulted from causes in both supply and demand sides. Banks were cautious to tie their resources long-term given the imposed requirements of prudential regulation and the risk involved in the existing non-transparent situation of their clients. Besides, they mostly lacked adequate infrastructure and instruments for generating long-term savings. At the same time, the barriers were on the demand side as well. Due to their accumulated liabilities and low creditworthiness, a considerable proportion of companies were neither in a position to effect investment nor creditworthy enough to be granted long-term credits. The mentioned bias in the term structure of extended credits follows from a comparison between the leading Czech bank, the Commercial Bank, and its German counterparts; however, it may be subject to various qualifications.

Table 3.5
Structure of credit and deposit markets in the Czech Republic (in %)

	Credit markets*			Deposit markets		
	12/92	6/93	12/93	12/92	6/93	12/93
Total large 'old' banks	68.4	67.3	64.5	88.8	83.9	78.9
Commercial Bank	31.5	31.3	29.6	25.9	22.0	22.9
Investment Bank	13.0	11.9	11.0	6.7	8.7	8.5
Czechoslovak Trade Bank	8.1	8.4	8.0	5.2	4.9	4.9
Czech Saving Bank	15.8	15.7	15.9	50.2	47.4	42.6
Total 'new' banks	17.5	20.8	24.1	10.3	15.3	19.9
Agrobank Prague	4.6	5.0	5.2	3.0	4.1	5.0
Bohemia Bank	0.8	1.1	1.3	0.8	1.6	1.7
Credit Bank Plzen	0.9	1.0	1.3	0.2	0.7	1.0
Moravia Bank	0.2	0.5	0.7	0.2	0.4	0.7
Post Bank	0.3	0.5	0.6	0.3	0.6	1.0
Consolidation Bank	14.1	11.9	11.4	1.7	1.7	1.2

Sectors' shares on total credits (in %)

Non-financial corporate sector	83.7	80.8	77.4	27.0	21.8	24.4
of which						
State companies	49.0	44.2	29.9	13.6	10.2	8.6
Private companies	32.0	34.1	44.3	11.4	10.1	13.6
Foreign controlled companies	2.6	2.4	3.2	2.0	1.5	2.2

Table 3.5 (continued)
Structure of credit and deposit markets in the Czech Republic (in %)

	Credit markets*			Deposit markets		
	12/92	6/93	12/93	12/92	6/93	12/93
	Sectors' shares on total credits (in %)					
Non-corporate private businesses	4.6	6.4	6.9	3.9	3.6	3.9
Households	8.0	7.1	6.9	46.3	47.4	44.7

*   banks' credits to enterprises and households in domestic currency
Source: Czech National Bank, own calculations.

Table 3.6
Maturity structure of bank credits* (shares in %)

	12/91	12/92	12/93	2/94
Short term	39.5	38.1	41.8	41.7
Medium term	18.2	27.2	28.4	28.4
Long term	42.3	34.7	29.8	29.9

*   bank credits to firms and households extended in domestic currency
Source: Czech National Bank, own calculations.

Structure of Czech banking and bad loans

In the Czech economy, as in other EMEs, accumulated bad loans originated under two different regimes: there were bad loans[7] extended prior to transition, i.e. a heritage of the past, and 'new' bad loans, only emerging in the course of transition. Despite some common grounds and causes, the ratio of 'old' to 'new' bad loans as well as their consequences differed across transition economies. And, accordingly, the policies adopted in individual countries also differed. While Hungary followed mostly a centralized approach to cope with the bad loans issue, in the Polish case a decentralized strategy was relied upon. The Czech policies made use of elements of both approaches.

Table 3.7
Maturity structure of credits and deposits in selected banks
(as of 31 December 1992, in %)

	Credits		Deposits		
	Short and medium term	*Long term beyond 4 years*	*Sight deposits*	*Term up to 4 years*	*Deposits beyond 4 years*
Berliner Bank	67.8	32.2	17.4	74.5	8.1
Commerzbank	63.9	36.1	22.0	59.6	18.4
Deutsche Bank	67.4	32.6	24.5	53.4	22.1
Commercial Bank	89.5	10.5	48.3	40.5	11.2

Source: Commercial Bank, own calculations.

Table 3.8 identifies risk credit's developments in the Czech Republic in the course of 1993 and compares them with the trends in banks' reserves and loan-loss provisions. It follows that the ratio of risk credit to total banks' credit deteriorated in 1993, suggesting prolongation of unfavorable longer-term trends. At the end of 1993 every fourth crown of extended credits was subject to the risk of not being repaid.[8] However, banks in the same period succeeded in substantially enlarging their reserves and provisions against losses. As a result, the coverage of risk credits improved, despite their absolute as well as relative increase. With reference to this it can be claimed that the stability of the banking system was enhanced in the discussed period. This conclusion applies to the banking sector as a whole, and not necessarily to individual banks or groups of banks.

Table 3.8
Risk credits, reserves and provisions of the banking institutions[a] in the Czech Republic

	12/92	*6/93*	*12/93*
Credits total[b]	567.8	615.0	672.3
Total risk (classified) credits	109.2	133.6	159.8
Percent share of risk credits on total credits	19.2	21.7	23.8
Reserves[c]	18.6	26.2	50.1
Loan-loss provisions	11.3	11.3	6.9
Percent share of reserves and provisions on risk credits	27.4	28.1	35.7

a   excluding the Central Bank
b   bank credits to firms and households extended in domestic currency, in billion CK
c   reserves against losses, not including bank reserve funds
Source: Czech National Bank, *Financial Statistical Information.*

Bad loans differed according to types of banks. Due to their differentiated starting endowments, operating conditions and behavioral standards, three groups of banks active in the Czech market continued to distinguish themselves: large (old) banks, small (new) banks and foreign banks.

Old banks, unlike new ones, could make use of the already established branch network and infrastructure for retail banking which provided ready access to primary deposits. At the same time, they benefited from accumulated expertise and information, however deficient. On the other hand, at the start of transition they all were undercapitalized and burdened with high shares of non-performing loans in their portfolios. The extent of these structural defects inherited from the past implied in fact their technical insolvency at the initial stage. Though such an insolvency was not likely to become real as far as there was an explicit or implicit state umbrella, they faced the requirement to comply with the targeted norms of prudential regulation and to work out the high share of bad loans. Accordingly, they pursued a fairly conservative approach in their credit policies. Besides, the authorities of former Czechoslovakia implemented earlier than in other transition countries a number of pragmatic measures targeted to at least relieve if not to solve the burden of the stock of bad loans inherited from the past. Those pragmatic steps included the establishment of the Consolidation Bank (March 1991), recapitalization of 'old' commercial banks (October 1991), several rounds of netting out companies' bilateral and multilateral arrears (for their discussion and evaluation see Hrncír, 1993). Though only partial, compromised and, moreover, delayed, these measures evidently alleviated the intensity of the bad loans problem of old banks.

As follows from table 3.9, the share of risk credits in total credits of large (old) banks did not increase in 1993. At the same time their reserves and loan loss provisions were substantially enlarged which resulted in a remarkable improvement of their ratio to risk credits. It follows that large banks seemed to be in much better shape at the end of 1993 compared to the start of transition.

The relative advantages and disadvantages of newly established domestic banks developed just the other way round compared to old banks. The new entrants were free from the burden of the past. On the other hand they mostly lacked branch networks in the initial period of their existence and, consequently, their opportunities to collect primary deposits were rather limited. They were much more dependent on Central Bank refinancing credits than large banks. While these credits amounted to only five to ten per cent of large banks' resources, they were as high as 60 per cent in the case of small banks. Confronted with the shortage of resources and/or their higher costs, some of the new banks appeared ready to compete by offering more favorable deposit rates which, in turn, implied higher lending rates. As a result, and in accordance with the concept of adverse selection, their clients and projects accepted for financing were biased to risky ones. Moreover, they had to deal with mostly newly emerging clients, lacking the previous creditworthiness record. The new entrants into banking services, being themselves as a rule inexperienced, were thus exposed to a high degree of uncertainty and risk.

When evaluating the outcomes, the available evidence seems to suggest a widely differentiated record. There are banks which within a relatively short time period acquired a high banking standard and fared possibly even better than old established banks. On the other hand, there were already cases of troubled banks and even failures.

As follows from table 3.9, the share of classified (risky) credits in the group of small domestic banks dramatically increased in the course of 1993. (Of course, the data also reflect the extension of external auditing and the registered increase is thus to some extent a statistical product). Nevertheless, at the end of 1993 the relative level of risk credits was still substantially lower than for big banks. However, because of much lower reserves and loan-loss provisions created, the group of small banks as a whole appeared much more fragile and sensitive to potential future pressures and losses.

Table 3.9
Risk credits and reserves across groups of banks

	Large banks[*]	Foreign banks	Small banks
	Share of risk credits on total credits (in %)		
December 1992	22.2	0.3	4.5
June 1993	23.0	0.7	10.5
December 1993	21.5	2.7	16.9
	Reserves and loan loss provisions (in billion CK)		
December 1992	22.1	0.3	1.0
June 1993	29.4	0.3	1.4
December 1993	45.5	0.6	2.9
	Share of reserves and loan loss provisions on risk credits (in %)		
December 1992	24.1	849.5	36.5
June 1993	28.2	247.4	16.2
December 1993	45.0	90.2	16.8

*   excluding Consolidation Bank
Source: Own calculations based on CNB data.

Private sector's share of bank credits

It has been often asserted (Phelps et al., 1993) that one of the main drawbacks, if not the single most damaging one, to banking intermediation in the EMEs is the continued flow of available funds to the wrong clients, in particular to the large, struggling state firms, at the expense of prospective clients especially of the emerging private sector. A number of reasons have been cited in this respect, including:

1   the continued symbiosis of state firms and their traditional creditors, exacerbated by the locked in effect on old banks;

2   the availability of collateral;

3   the excessively high level of interest rates constraining new entrants in particular;

4   the lack of expertise in assessing the creditworthiness and the degree of risk of individual clients and projects.

The main alleged culprit, the symbiotic relationship between the old banks and state enterprises, was in the past hardly avoidable. This was especially true for ex-Czechoslovakia, where there was virtually no significant private sector as late as 1990. However, as follows from table 3.10, a remarkably increased volume of bank credits was earmarked for the private sector since then. At the end of 1993 the share of the private sector (including cooperatives) of total bank credits in the Czech Republic already surpassed 50 per cent, compared to 10 per cent at the end of 1991 (plus a further 10 per cent allocated at that time to cooperatives, which have since mostly been transformed).

The available data for 1994 seem to suggest the shift in favor of the private sector continued, if not further accelerated. It must be noted, however, that the presented data do not distinguish between the newly-emerging private businesses, firms sold to domestic and foreign investors and firms privatized through the voucher scheme. Accordingly, the resulting proportions in the sectors' shares may appear less persuasive. Nevertheless, the evidence on new credits extended by individual banks supports the conclusion that a significant reallocation of credits has been recently going on in favor of the private sector.[9]

The Czech case may be specific in the sense that most old banks developed their activities only with the start of transition and shortly afterwards they were privatized within the framework of the voucher scheme. These old banks therefore were not much older than the new small private ones. Besides, starting their activities in a reform environment, with changed incentives and with mostly newly recruited management, the behavior pattern, policies and preferences of old banks in principle have not diverged from those of new banks. This characteristic may be one of the reasons why there was no evident bias in their lending in favor of traditional clients.

Table 3.10
Sector's shares on bank credits in the Czech Republic

	Dec. 1991		Dec. 1992		Dec. 1993		Feb. 1994	
	bn. CK	%	bn. CK	%	bn. CK	%	bn. CK	%
State firms	333.7	66.89	283.7	47.71	200.7	28.8	165.0	23.3
Cooperatives[a]	49.9	10.00	45.9	7.72	-	-	-	-
Private sector	48.8	9.78	166.6	28.02	350.6	50.3	393.0	55.6
Foreign exchange credits to residents	3.5	0.70	16.0	2.69	24.2	3.5	26.0	3.7
Households	36.3	7.28	46.5	7.82	46.5	6.7	45.5	6.4
Other[b]	26.7	5.35	35.9	6.04	74.5	10.7	77.5	11.0
Total credits	498.9	100.00	594.6	100.00	696.5	100.0	707.0	100.0

a   since the beginning of 1993 the credits extended to cooperatives are included in the private sector's share

b   other are mostly unidentified cases, in the process of change in their status, i.e. especially where privatization was underway

Source: Own calculations based on CNB data.

Conclusion: banks and corporate governance

The Law on Banks (no. 21/1992) provided for a universal type of banking, integrating both commercial and investment banking activities. The discussion of the advantages and constraints of universal banks versus separated commercial and investment banking is related to the strategic issue: what should the role of banks be in the 'real' sector and in corporate governance in the EMEs?

Though not necessarily implying the German model of the banks' role in the non-financial sector, the universal bank does create prerequisites for such an approach. Banks, while extending credits and holding equity at the same time, are in a better position, due to intimate insider information, to manage the developments of the firm, exercise control over projects and investments and, having interest in long-term profitability, to smooth the periods of financial distress. For the specific conditions of EMEs, it seems to be of particular importance that universal banks could have a strong position to enforce early restructuring of enterprises, short of a bankruptcy situation.[10]

However, there are at least two types of potential constraints. The first is related to the risk inherent in the universal banking solution as such. The arguments discussed in the literature include moral hazard problems, potential conflict of interest, and the monopoly position of conglomerate financial institutions in the economy (Saunders and Walter, 1991). To cope with such risks involved, the EMEs would have to further improve the regulatory system, anti-monopoly procedures, banks' reputational standard and to continue in opening the financial system to the new entries and competition from abroad (Nuti, 1989).

The second type of constraint is related to the specific conditions in the EMEs. On the one hand, the given model could imply the risk of maintaining and even promoting the 'traditional' behavior of keeping non-viable firms afloat. On the other, it was not a priori evident why banks in transition economies should be interested in such a solution and, besides, whether they are also capable of implementing it.

Though with some limits, the introduced legal framework in the Czech Republic provided for the banks to hold equities of non-banking companies.[11] These limits were not as yet, however, effective limits. Rather, banks were discouraged by the non-transparent situation of companies and by the unsettled, fluctuating ratios of share prices. In the circumstances, the risk of trading bad debt for bad equity was relatively high. Besides, as co-owners of firms which go bankrupt, banks are in a much less favorable position vis-à-vis other creditors in recovering their assets.

The important role in the entire process is held by the Investment Privatization Funds (IPFs). They developed spontaneously, resolved the potential low participation problem and became block holders, correcting for the threat of excessively wide dispersion of shares in voucher privatization. The funds, founded by banks and other financial institutions, were particularly successful in attracting a majority of investors and became owners of a substantial share of property privatized in the first wave of the voucher scheme implemented in 1992-93. About 72 per cent of 'points' have been assigned to IPFs, of which ten top funds collected more than a half (i.e. about 40 per cent of all voucher-points in the first wave). From these ten top funds nine were founded by commercial banks and other financial institutions.

Though relatively lower in the second wave of voucher privatization started in 1994, the share of investment funds and mutual funds in the privatized property continued to be dominant. About 65 per cent of voucher points were invested through funds in the second wave compared to 72 per cent in the first wave.

Due to their share in privatization property, the behavior, preferences and activities of investment funds, especially of those which have attracted substantial portions of voucher-points, are of crucial importance. The distribution of shares which developed from the first wave was mostly considered as more or less optimal. Both the potential extremes, i.e. too dispersed or overly concentrated ownership did not happen. Nevertheless, as could be expected, the available accumulated evidence suggested considerable differentiation among investment funds. Though their number increased in the second wave to 349, the prospects of survival for a significant share are bleak. The process, including mergers, was already underway. As late as in the first half of 1994 some of the existing funds also have not complied with the rules under which they are obliged to provide investors with adequate information on their financial performance and to issue shares to their subscribers.

A strategic, still unresolved issue was the role of investment funds in corporate governance and the implied ways of affecting corporate performance and behavior. Both alternative options, investment funds confined to managing the portfolio of shares and investment funds as active proprietors, directly engaged in exercising ownership functions, were still possible. The existing legal framework provided, with some restrictions,[12] for both options. The available evidence seemed to suggest a wide divergence of behavior across individual investment funds as yet and a parallel divergence applied to the views held in the ongoing discussion on the subject.

However desirable an active and engaged proprietor's function of investment funds may be, constraining factors were in place. A number of funds acquired a rather wide portfolio in the first privatization wave, which could hardly be manageable in an active overseeing of the respective companies. At the same time, the conflict between the short-term interests and liquidity pressures may increasingly interfere with the intention to act as long-run investors. Moreover, the capabilities of individual funds to contribute to the companies' restructuring remained to be proved. There were apparent differences in the standard of their 'human capital' and, accordingly, in the concepts and policies which their representatives may advocate in the companies due to their former positions and experience.

Notes

1. This is a revised version of the paper presented at the ACE Conference on Company Management and Capital Markets, Prague, 24-26 April 1993. For helpful comments on an earlier draft I am particularly indebted to Marvin Jackson, Stavros Thomadakis and Jan Klacek. I would also like to thank the participants of the Prague Conference for stimulating discussion.
2. Both Hungary and Poland effected this change earlier, in 1987 and 1988 respectively.
3. In Poland, high inflation rates of 1989 and 1990 effectively eroded real value of the former enterprise debts. This may explain the fact that a substantial deterioration of capital/asset ratio was reported to have taken place only afterwards (OECD, 1992).
4. The data on inter-enterprise indebtedness were based on two different sources: (a) bank data were only available until March 1991. Since then, with the changes in the payments regime, banks have not been obliged any longer to accept bills for payments against an enterprise with insufficient funds in its account with the particular bank. As a result, the bank data ceased to reflect the entire volume of payments in arrears; (b) since the first quarter of 1991 until September 1991, the available official sources are enterprise data as collected by the Ministry of Finance. After this date the official institutions, including the Statistical Office as well as the banks, discontinued the disclosure of available evidence on enterprise indebtedness. All data referring to later periods are only based on estimates, sample surveys and expert evaluations.
5. For data on inter-enterprise arrears in former Czechoslovakia during the period 1982-90 see Hrncír and Klacek (1991).
6. The data represent the cumulative amounts of liabilities of firms in payments arrears whereas the asset counterpart is not identified. Hence the respective amounts of outstanding liabilities do not imply the corresponding overhang of liabilities of the Czechoslovak economy as a whole vis-à-vis the rest of the world.
7. In the discussion on the EMEs this concept was often used in an ambiguous way, lacking identification. Two different interpretations are in principle possible, wide and narrow ones. While the former comprises all types of qualified, non-standard loans (the Czech banking statistics include them under the heading of risk credits), in the latter case it is confined only to the 'worst' sub-category of qualified loans, i.e. to non-performing ones. The latter interpretation is adopted in this paper.
8. Such a share, i.e. in the range of one fifth to one fourth of total bank credits, used to be referred to, especially by external observers, as the figure representing the volume of non-performing loans in former Czechoslovakia. Blommenstein and Lange (1993), for example, in the recent OECD study quoted 21 per cent. Our paper, in accordance with the approach of Czech banking statistics, refers to bad, non-performing loans as to only one, the 'worst' sub-category of classified loans.

The figures in table 3.7, amounting to the level of 19.2 to 23.8 per cent of total banks' credit for 1993, represent all types of classified loans.

9. Begg and Portes cited the evidence from the survey by L. Webster (World Bank) on the access of private manufacturing firms to bank credit and concluded:

> The CSFR has very high private sector dealings with the banking system. 80 per cent of respondents has a bank loan, many had several loans. To date, new private business borrowing has perhaps been impeded less in the CSFR than in Poland, and especially Hungary (Begg and Portes, 1992, p. 16).

10. Taking into account present-day conditions in the East European region, the paper argues that one class of mechanisms, namely, outsider control by banks and other financial intermediaries, is well-designed to promote enterprise performance, while some of the other mechanisms, such as stock market or foreign investment, will not be strong enough in the near future, if ever, to be a major source of outsider governance. (Phelps et al., 1993, p. 38).

11. In accordance with the Law on Banks, commercial banks were entitled to acquire shares up to ten per cent of capital of a non-banking institution. The total volume of their stakes in non-banking firms should not, however, exceed 25 per cent of their own capital and reserves. To hold stakes exceeding these norms, the prior consent of the Central Bank was required.

12. According to the legal rules in force, investment funds were prevented, for example, from holding more than a 20 per cent share in any single company.

References

Abel, I. and Bonin, J. P. (1993), *Financial sector reform in the economies in transition: on the way to privatizing commercial banks*, Conference Paper, Vienna.

Bank for International Settlements (1993), *63rd Annual report*, Basle.

Begg, D. and Portes, R. (1992), 'Enterprise debt and economic transformation in Central and Eastern Europe', *CEPR Discussion Paper*, no. 695.

Blommenstein, H. J. and Lange, J. R. (1993), 'Balance sheet restructuring and privatization of the banks' in OECD, *Transforming of the banking system: portfolio restructuring, privatization and the payment system*, Paris.

Corbett, J. and Mayer, C. (1991), 'Financial reform in Eastern Europe: progress with the wrong model', *Oxford Review of Economic Policy*, vol. 7, no. 4, pp. 57-75.

Dittus, P. (1993), 'Finance and corporate governance in Eastern Europe' in Fischer, B. (ed.), *Investment and financing in developing countries*, Nomos Verlagsgesellschaft, Baden-Baden.

Hare, P. (1991), 'The assessment: microeconomics of transition in Eastern Europe, *Oxford Review of Economic Policy*, vol. 7, no. 4.

Hrncír, M. (1993), 'Financial intermediation in former Czechoslovakia and in the Czech Republic: lessons and progress evaluation', *Economic Systems*, vol. 17, no. 4, pp. 301-327.

Hrncír, M. and Klacek, J. (1991), 'Stabilization policies and currency convertibility in Czechoslovakia', *European Economy*, Special edition, no. 2, pp. 17-39.

Jackson, M. (1994), *Property rights, ownership, company organization and governance in the transition*, this volume.

Nuti, D. M. (1989), 'Feasible financial innovation under market socialism' in Kessides, Ch., King, T., Nuti, M., and Sokil, C. (eds), *Financial reform in socialist economies*, World Bank, Washington.

Nyberg, P. and Vihriäla, V. (1993), 'The Finnish banking crisis and its handling', *Discussion Papers*, no. 15, Bank of Finland, Helsinki.

OECD (1992), *Economic Survey of Poland*, Paris.

Phelps, E.S., Frydman, R., Rapaczynski A. and Shleifer, A. (1993), 'Needed mechanism of corporate governance and finance in Eastern Europe', *EBRD Working Paper*, no. 1.

Ptácek, J. (1993), 'Insolvency of the part of the Czech enterprises and the possibilities of its solution, (in Czech), *Research Report*, no. 9, Institute of Economics, Czech National Bank, Prague.

Saunders A. and Walter I. (1991), 'Reconfiguration of banking and capital markets in Eastern Europe', *The Journal of International Securities Markets*, Autumn, pp. 221-238.

Schmieding H., and Buch C. (1992), 'Better banks for Eastern Europe', *Discussion Paper*, no. 197, Kiel.

Thomadakis, S. (1994), *Perspectives on financial system and enterprise efficiency in the transition economies of Central and Southeastern Europe*, this volume.

4 Financial structure, performance and the banks

Alena Buchtíková and Ales Capek[1]

Introduction

After all the basic systemic reform steps have been undertaken it is now time for the microeconomic restructuring and consolidation processes to fulfill expectations and to launch the process of economic growth. Privatization was expected to create the necessary framework for this restructuring. Privatization, however, cannot implement this task by itself. At best it will create new, more efficient ownership structures and lead to some organizational changes. Real restructuring based on changes in industrial output and the underlying technologies as well as commercial restructuring including the build up of new sales networks is, however, a more complicated, time consuming and, what is perhaps most important, financially very extensive task.

The quantitative dimension, the amount of money needed for restructuring is not the only issue. From a certain point of view even more important are the ways in which this finance is provided, and within what financial system. The system of finance of corporate growth has implications for the structure of corporate governance and thus affects the efficiency of the management of firms. The importance of this aspect in the Czech Republic is increased by the fact that in most cases privatization was not accompanied by any inflow of finance to the firms. One consequence of this situation is that post privatization restructuring will create a great demand for external finance in enterprises. The second consequence is that the 'corporate governance outcome' of the privatization process is, in many cases, very preliminary and it will be gradually shaped by the evolving financial structure of the enterprises. The commercial banks will play a decisive role in this process as financial intermediaries providing the needed funding but also as subjects involved in the principal-agent relation vis-à-vis the corporate sector both as the creditors or via the investment companies which they have created for privatization purposes.

83

In this chapter we will provide some theoretical as well as empirical background for the discussion about the financial structure of enterprises and the microeconomic role of banks.

Financial structure and economic performance

Economic literature now generally accepts the view that financial structure of the firm matters, in that the different ways in which enterprise investment is financed affect the value of the firm.[2] The reason for this is that the firm is understood not as a profit maximizing production function in an information costs and moral hazard free world, but as an economic entity comprised of economic agents with different behavioral priorities, whose economic relations within the firm are regulated on the basis of incomplete contracts. The way in which the firm is financed determines the incentive framework established within the firm.

There are three basic issues which are of critical importance. The first is the financial structure in terms of internal financial sources and external finance which may have the form of debt or equity. The second is the level of concentration of external finance. And the third is whether external finance is provided by the banking system or via capital markets.

Financial structure of the firm: inside and outside finance

The simplest case is when the firm is financed completely from internal sources. This is at the same time the most efficient arrangement from the point of view of managerial incentives. Assuming that the owner's priority is the maximum value of the firm, the owner-manager in one person follows a straightforward maximizing strategy. There are no conflicts between his behavior as a manager and his priorities as an owner and consequently no agency costs arise on this level. There are of course agency costs within the firm on all lower levels of principal-agent relationships within the enterprise.[3] It is, however, rare that the firm would be completely financed from internal sources. External finance becomes inevitable in general in situations where there are attractive investment projects available for the firm but not enough internal sources for their funding. Especially for the most dynamic growing firms at some moment external finance becomes a necessity. As the firm matures and its financial strength increases, it might be capable of financing its investments largely from internal sources but even the big firms issue debt and equity to finance their growth. Perhaps the difference is that, in the case of the biggest firms, rather than absolute lack of internal finance, the opportunity costs with which the external funds are raised play a more important role than the cause of external finance.

The life-cycle financial development of the firm provides one explanation why during different phases of the growth of the firm its financial structure changes.

A slightly different look at the evolution of the composition of sources of corporate finance is one which takes into account structural changes in the economy. Assuming that the economy moves in long-term Schumpeterian technology based cycles with periods of industrial and organizational restructuring followed by the spread and maturation of the new industrial structure, we could expect certain cyclical movement in corporate finance as well. It would be reasonable to expect that external finance will be more important in initial phases of industrial restructuring when new investment opportunities are rapidly emerging, when many new firms enter the scene and the old ones undergo costly reshaping. Schumpeter (1955), recognized this aspect when he stressed the importance of credit for the innovation activity of the entrepreneur.

The system of financing enterprises is, in the above approach, a historical phenomenon and the financial systems of different countries are affected by the history of their industrial development. Gerschenkron (1962) in this context explains the rapid development of the German bank based financial system at the end of 19th century by the needs of German industrialization in a situation of economic backwardness and lack of internally accumulated funds. The banks under these circumstances became the suppliers of both the needed finance as well as the entrepreneurial initiative.

It should be stressed, however, that the evolution of the financial system is not shaped only by the phases of industrial cycle but as well by the type of emerging technico-economic structure. As Sabel and Piore (1984) have indicated, the new type of technology which is the basis of current structural changes in the developed economies is very different from the technology typical for postwar industrial growth. The new pattern of industrial development characterized by accelerated innovation, entrepreneurial dynamism and shifts in market structures requires the adjustment of the financial system which in the 60s and 70s served very different clients. The big firms of the past were in many cases and to a great extent independent of external finance. However, due to their long-term economic record lowering the information costs of external finance, they were (and are) capable of raising considerable amounts of external funds. The new, 'unknown' but very dynamic firms are in a different position. They are in need of external funds, they promise great returns on borrowed funds but as well are subject to considerable risk. The management of risk is thus becoming one of the more distinct features of current financial systems.

All the above remarks point to one conclusion: the financial system (i.e. for us the system of financing the firms) is not something which is given to the economy from above. It reflects the phase and type of technico-economic development, as well as some institutional peculiarities in a given country, and permanently adjusts to the new situation. The issue is not then necessarily to take-over the German or English financial system but to assist in the development of a financial system relevant to the restructuring needs of the economy which is at the same time determined by the historically established financial structure and the institutional framework so far created during economic reform.

Let's now look a little bit more closely at the characteristics of different types of external finance. We have already mentioned the advantage of inside finance resulting from the absence of agency problems. We have as well mentioned that internal funds may not be available each time they are needed. The incentive and information problems of external finance on the one hand and the availability of agency cost free inside funding on the other thus create the general framework within which the concrete pattern of the financial structure of the firm emerges.

Table 4.1
Sources of finance of physical investment in selected countries between 1970-89 (%)

	U.K.	U.S.	Germany	Japan	France
Internal	97.3	91.3	80.6	69.3	65.8
Bank finance	19.5	16.6	11.0	30.5	27.8
Bonds	3.5	17.1	-0.6	4.7	1.5
New equity	-10.4	-8.8	0.9	3.7	6.3
Trade credit	-1.4	-3.7	-1.9	-8.1	-0.5
Capital transfers	2.5	-	8.5	-0.1	2.5
Other	-2.9	-3.8	1.5	-	0.4
Statistical adj.	-8.0	-8.7	0.0	0.0	-4.6

Source: Edwards. J. (1993), *The financing of industry, 1970-89: an international comparison,* unpublished manuscript.

Empirical studies indicate substantial reliance on internal funds when financing the growth of firms in developed market economies. Edwards (1993) shows that in the main market economies, internal sources clearly dominate the pattern of financing corporate investment. The above numbers are a little bit surprising taking into account all the talk about the importance of capital markets in industrial development. We should nevertheless be aware of the fact that almost all of the trading with corporate securities represents secondary trading of corporate debt and equity and that the huge amounts of securities traded on capital markets are in no direct relation to physical investments of the enterprises.

Equity and debt

The two principal forms of external finance are debt and equity. Each of them creates a slightly different incentive framework for the management of the firm.

Let's assume first that the firm is 100 per cent financed internally and owned by the manager. He maximizes his utility function consisting of pecuniary and nonpecuniary benefits and in equilibrium, marginal utility of each of these 'benefit' items related to additional expenditure must be equal.

Then for lack of internal sources of financing for the growth of the firm the manager decides to draw on external funds in the form of equity. As a consequence of this change in the financial structure of the firm, his priorities will change. Assuming that he is still able to enjoy all the nonpecuniary benefits but only a share of pecuniary benefits proportional to his share in the firm, he will be quite naturally less motivated to expend resources with the aim of increasing the value of the firm and more inclined to use them for achieving the nonpecuniary gains. The interests of the inside and outside owners are thus becoming diverse and the agency problem arises. Monitoring of the managers' activities is one way to constrain them in the above behavior. The rising monitoring costs will, however, be taken into account by the capital markets and the result will be the decline in the value of the firm.

The incentives to monitor the managers depend to a great extent on the level of concentration of external owners. Monitoring becomes reasonable in a situation of concentrated external ownership when the dominant external shareholder will appropriate most of the benefits resulting from monitoring. If, on the other hand, external ownership is dispersed, no individual shareholder will be, due to the free rider problem, interested in bearing the full costs of monitoring and gain only his proportional share of monitoring benefits.

Investing through financial intermediaries may be a solution to this problem as the financial intermediaries, due to their ownership strength and professional approach, represent more suitable monitors. Diamond (1984) has developed a model of financial intermediation as delegated monitoring and even though he focused on a case where the intermediary provides the entrepreneur with a loan and not equity capital, some of his conclusions are of general importance, namely that the diversification of the intermediary's investment portfolio brings with itself the information advantage and decreases the monitoring costs per entrepreneur or that the optimal relation between the intermediary and its depositors is the debt contract due to its incentive effect on the intermediary. We shall talk about the incentive effect of the debt arrangement later.

Another mechanism putting constraint on the behavior of managers is the possibility of take-over in cases when the value of a poorly managed firm decreases considerably below its potential value under alternative management. The extensive literature about take-overs, however, mentions some problems of this procedure. One of them is that ascertaining the potential value of the firm is, due to the information asymmetries, a costly as well as a risky task. Another is that the current owners, under assumption of rational behavior, would free ride on the information gathered by the raider and never sell the stocks for less than their potential value in which case there would be no reason for the raider to continue in bidding (see Grossman and Hart, 1980). The knowledge of these problems connected with take-overs by the managers gives them some freedom as to their value of the firm's non maximizing behavior.

The good news about outside equity, on the other hand, is that it enables one to spread the risk of managerial decisions between owners-managers and outside equity holders. Unlike with debt, the decline in firm returns will be borne both by the firm and its external shareholders as both the internal and external owners are the residual claimants on firm profits.

The basic difference of external finance by debt is that the payments by the firm to outside financiers are fixed and in principle cannot be changed. This fact is very important as concerns the incentive framework created by debt finance. Because the payments to creditors are fixed whatever the economic returns of the firm, the managers are becoming the residual claimants within the firm and thus are motivated to devote higher levels of effort focused on increases in the value of the firm compared with the situation of being financed by equity, when they had to share the increases in profits with the stockholders. In other words, the owners-managers bear within the debt arrangement full consequences of their actions and thus are motivated to socially optimal behavior.

This 'incentive effect' of the debt contract is beneficial both for the creditors and the managers. For the creditors the debt arrangement is an instrument that automatically motivates the managers, and thus lowers the monitoring costs. The managers, on the other hand, besides receiving the total residual income are, except for situations of bankruptcy, protected from interference from those who provide the external finance. The creditors do not have any right to govern the management of the firm or even change it.

The above characteristic of the debt contract is of course very simplistic and in reality the balance of its costs and benefits differs according to the concrete circumstances. Two things seem to be of importance: the relation between the net worth of the investing firm and the amount of external finance by debt and the proximity of bankruptcy.

Let us start with the first issue. One specific situation which modifies the incentive effect of the debt contract is the situation of financial fragility. (Bernanke and Gertler, 1990). A situation of financial fragility occurs when entrepreneurs eager to pursue their investment projects have low net worth and rely heavily on external finance by debt. Even though they 'bear the full consequences of their actions' the small amount of their own wealth which they bring to the project makes them undertake actions which are not desirable from the point of view of the creditors, e.g. to take on investment projects which are too risky. Jensen and Meckling (1976, p. 344) illustrate this by the analogy of a poker player playing on money borrowed at a fixed interest rate, whose own liability is limited to a very small stake. This will adversely affect the dynamics of investment activity in at least two ways. The lenders who realize the above dangers of financial fragility will be generally reluctant to lend funds to low net worth enterprises or at least they will offer their funds for higher interest rates which will constrain investment. Second, the information asymmetry between the lenders and borrowers creates the information problem and the lenders who are willing to lend their money have to bear the agency costs associated with their efforts to ascertain the conditions of the project. As a result the prospective costs of investment finance increase and consequently its amount decreases.

Financial fragility characterizes phases of initial economic development or extensive structural changes in the economy when there emerge numerous investment projects and the firms which are supposed to realize them have low net worth. Bernanke and Gertler (1990) offer a very straightforward but rather mechanistic cure to the problem: direct transfers of wealth to the potential borrowers to increase their net worth.

The second specificity of the debt contract concerns bankruptcy and the costs of it. We have mentioned above that the debt arrangement through its incentive effect encourages managers to adopt socially optimal measures. This is generally valid in situations when the firm is in good financial shape. When the financial health of the firm deteriorates the costs of the debt contract become more transparent. Economic literature differentiates between different types of agency costs connected with bankruptcy. One category is the bankruptcy and reorganization costs to be borne in situations when the firm really goes bankrupt. A specific agency problem also emerges, however, in situations when the firm is more or less close to bankruptcy. When the firm is doing poorly but bankruptcy so far is not the immediate alternative the managers might deviate from their socially optimal behavior and become quite conservative in their decision-making (see Bernanke and Campbell, 1988, p. 93). Instead of pursuing promising investment opportunities they might hesitate to undertake any dramatic entrepreneurial actions so as to avoid potential errors which could shift the firm closer to bankruptcy. The result is suboptimal choice of risk and this strategy may unfortunately lead to further deterioration of the financial situation of the firm.

When the firm is, on the other hand, already very close to bankruptcy or bankruptcy is almost certain, and moreover the firm is financed mainly by debt, the managers might incline to undertake even very risky projects. They do not have much to lose and thus they might choose a strategy close to gambling.

The problem with all these deviations from socially optimal managerial behavior is not only that they happen, that they lead to agency problems and affect the value of the firms but as well that rational lenders expecting the above types of behavior will be reluctant to provide the firms with external finance which will then suffer from a shortage of funds to support investment activity.

The above cost-benefit approach to different methods of external finance has illustrated the basic difference in managerial behavior under the two financial regimes. Financing by external equity leads to suboptimal profit maximizing effort and optimal choice of riskiness by the managers, a result caused by the sharing of achieved profits between the managers and external shareholders. The debt contract, on the other hand, provides the appropriate incentives for profit maximization but a faulty incentive framework for the choice of risk.

Intuitively some combination of external finance by debt and equity could create a 'compromise' incentive framework for the firm which would be preferable to the extreme situations of financing either by debt or equity and which would combine their benefits and/or reduce their costs. Jensen and Meckling (1976) have presented an introductory approach to this problem arguing that the optimal ratio of outside equity to debt is one which minimizes the sum of agency costs connected with both types of external finance. Jensen and Meckling's assumption is that the marginal agency costs behave in a standard way, i.e. they are increasing with increasing shares of the given type of external finance and thus the optimum structure of external finance must combine some non zero shares of both debt and equity. Optimum means that, given the extent of external finance, the minimum agency costs due to the 'efficient' mix of debt and equity lead to maximum value of the firm. In other words the value of debt and equity as determined in the capital markets reflects the agency costs attributable to the given external debt equity ratio.

The economic literature does not say, however, what exactly the ratio should be. The empirical analysis, on the other hand, indicates that the financial structure of firms in spite of some differences between industries, is fairly standard. One empirical survey for manufacturing in the U.S. in 1986 shows that the debt ratio (long-term debt incl. leases/long-term debt + equity) fluctuates between roughly 30 and 50 per cent (United States Department of Commerce, 1986).

So far we have focused on the financial structure of the firm and its role in creating the incentive framework for the management. Though it looks somewhat simplified, we should realize that any specific incentive tool used to motivate managers' optimal behavior is in some way connected with the financial structure of the firm. Phelps, Frydman, Rapaczynski and Shleifer (1993), enumerate eight principal incentive devices for managers and all of them are in some way related to the ways in which the firm is financed. Mayhew and Seabright (1992), similarly see the motivation framework for the management of firms being shaped mainly by the debt and equity holders.

We should, however, make one additional remark. In many or perhaps most cases the managers are not automatically ex ante the holders of equity in their enterprises, but ordinary professionals hired by the owners. Thus they do not qualify automatically as the residual claimants on the enterprise profits which of course would modify the incentive framework. The solution is the creation of such remuneration schemes for managers which would simulate their position as equity holders and at the same time create an incentive structure leading to their optimal choice of effort and risk. The problem is analogous to the emergence of optimal financial structure of the firm. The remuneration scheme for the managers should on the one hand make them feel the results of their actions but on the other hand protect them from unforeseen developments. The optimal structure of their remuneration should thus be comprised of a fixed part and a flexible part reflecting the increases in the value of the firm such as shares in profits, equity stakes in the firm, and so on.

Financial performance of Czech industry

Debt

As indicated above, almost all investment in western countries is financed from internal sources and the rest mainly by debt in the form of bank credit. In a reforming economy we would perhaps expect the reverse, i.e. that due to the poor performance of the enterprises and their low profitability, their investment will be financed to a great extent from external sources. In reality, however, the pattern of financing of enterprise investment in the Czech economy is not very far from that observed in developed countries. According to official statistics, in the first half of 1993, in the nonfinancial corporate sector 72.7 per cent of investment was financed by internal sources, 16.8 per cent by bank credit, 4.8 per cent by government subsidies, 2.0 per cent from abroad and 3.7 per cent by other sources (Hájek et al., 1993, p. 47). The structure of sources for investment in the 1980s is not very far from the above pattern. According to the Czechoslovak Statistical Office data during 1980-91, 45 to 60 per cent of enterprise investment in the Czechoslovak economy was financed from depreciation, 16 to 25 per cent from retained profits, 17 to 22 per cent from investment credits, 8 to 11 per cent from subsidies and 0.5 to 1.5 per cent from other sources. This illustrates that the enterprise sector was to a great extent capable of financing its investment needs.

Let us take furthermore into account that during 1992 the corporate sector's savings at the banks increased by about CK 24 billion which, in the absence of other financial instruments, can be understood as investments into financial assets and that at the same time, somewhat surprisingly, the total liabilities of enterprises with other firms declined by more than CK 60 billion (according to the Czech Statistical Office) - i.e. the firms were able to repay their debts in this amount. The enterprise sector thus created a volume of funds which more than covered its investment needs. This is in contrast to the general belief about the lack of resources for financing the growth of the firms.

Perhaps in 1994 in the postprivatization period, when a massive restructuring effort is expected, this situation will change and external finance will play a more important role. So far, however, the data do not indicate that a shortage of internal resources would decisively constrain investment activity.

The above numbers about the share of bank credit in financing enterprise investment, on the other hand, do not say very much about the importance of total enterprise debt including the total stock of accumulated bank credits from the past as well as debt due to inter-enterprise trade credits and the macroeconomic implications of this total debt.

Let us look therefore at more empirical data which will illustrate the financial structure of Czech enterprises. For the calculations in table 4.2 we have used the database of the Ministry of Industry and Trade covering 1,325 state-owned industrial enterprises in 1992. We have calculated the following indicators using the book values: LDE, BCD, LDA and DA.

The last two indicators, the LDA and DA ratios, illustrate the level of financial fragility of enterprises across industries. Their values show what share of assets of the firms is financed by debt. The debt-equity ratio (LDE) is the traditional leverage indicator describing the financial structure of the firm.

The DA indicator is highest within the machinery and electrotechnical industry and lowest within construction, construction materials and glass, china and ceramics. The LDA indicator and the debt-equity ratio are highest within the machinery, pulp and paper and clothing industries and lowest within construction and construction materials, glass, china and ceramics as well as iron and steel and leather, furs and footwear.

What is the importance of these indicators and what is the danger of financial fragility? We have shown already above that financial fragility is caused by high leverage. Bernanke and Campbell (1988) argue that financially fragile firms are very easily vulnerable to external shocks which further lower their net worth. Such firms find it more costly and difficult to raise external funds to finance their investments as well as current production. As a result the recession triggered by an initial negative shock in an economy populated by highly indebted firms will be much deeper and the interest rates higher compared with a less leveraged economy.

Table 4.2
Indicators of financial structure in Czech industry in 1992

Industry	LDE	BCD	LDA	DA	LDE var. coef.
Total	0.22	0.53	0.12	0.35	1
Iron and steel	0.14	0.45	0.09	0.33	0.01
Non-ferrous metallurgy	0.18	0.54	0.11	0.35	0.01
Chemicals and rubber	0.18	0.45	0.12	0.34	0.03
Engineering	0.34	0.57	0.15	0.40	0.07
Electrotechnical	0.22	0.50	0.12	0.43	0.03
Construction materials	0.10	0.65	0.08	0.17	0.08
Wood processing	0.18	0.61	0.12	0.34	5.18
Metal processing	0.16	0.58	0.10	0.37	0.02
Pulp and paper	0.25	0.59	0.16	0.34	0.11
Glass, china and ceramics	0.13	0.62	0.08	0.21	0.04
Textiles	0.16	0.57	0.10	0.36	0.04
Clothing	0.24	0.67	0.15	0.36	0.01
Leather, footwear and furs	0.14	0.59	0.09	0.36	0.50
Printing	0.21	0.56	0.13	0.34	0.04
Other industrial activities	0.20	0.61	0.13	0.35	0.81
Building and civil engineer.	0.14	0.44	0.05	0.19	0.16

LDE   long-term debts/equity
BCD   bank credit/total liabilities
LDA   long-term debts/assets
DA   total liabilities/assets
Source: Czech Ministry of Industry and Trade database, own calculations.

High leverage may become a problem especially in periods of wide fluctuations in the capital markets. Davis (1992, p. 180) shows by using historical data that the market values of share prices during financial crises move sometimes by tens of per cent. It is mainly the financially fragile firms which are endangered during these periods. In the highly leveraged firms with high DA, LDA or LDE ratios there is even the danger that in case of the decline of the market value of assets the company's debts may easily exceed the company's assets, the net worth of the firm becomes negative and the firm becomes insolvent. In the Czech economy where high fluctuations in the share prices are due to the embryonic stage of development of the capital market's 'daily' experience and where on the other hand the market value of companies' net worth is quite low, financial fragility is a major problem.

The share of bank credit in total liabilities, as expressed by the BCD indicator, lies between 44 per cent in construction and 67 per cent in the clothing industry. On average roughly one half of total industry debt is owed to the banking sector. We have as well calculated the variation coefficient for the debt equity ratio which indicates that in most industries except wood processing, leather, furs and footwear and other industries, the leverage is quite stable across all firms within given industry.

Table 4.3
Indicators of financial structure in different size categories in 1992

	Number of firms	LDE	BCD	LDA	DA
Total	1,050	0.22	0.53	0.12	0.35
1st quartile	biggest 50	0.27	0.54	0.14	0.37
	212	0.20	0.53	0.11	0.35
2nd quartile	262	0.17	0.53	0.09	0.32
3rd quartile	262	0.18	0.51	0.10	0.30
4th quartile	264	0.15	0.49	0.09	0.28

Source: Czech Ministry of Industry and Trade database, own calculations.

The pattern of financial structure within the size distribution of firms as illustrated by table 4.3 is quite clear. Leverage is increasing with the size of the firms. We can speculate that the big firms have better access to credits, that due to their size they are better clients for the banks or simply that banks might feel more endangered by refusing to provide large enterprises with credits. At the same time and for similar reasons the large firms might more easily get indebted among themselves. The firms are more willing to credit their big customers than small or episodic customers.

The total size of enterprise debt does not tell the whole story about enterprise indebtedness. In the Czech Republic as well as in other reforming economies, bad debts comprising both the inter-enterprise arrears and the bad assets in the balances of the banks vis-à-vis the industrial sector are a characteristic phenomenon. Table 4.4 contains some indicators of the bad debt problem. The following ratios were calculated: TBDS, PBDS, TBDD, TBDE and TBDA.

Bad debt represented a little bit more than 20 per cent of total enterprise debts in 1992 with differences among industries. The data indicate that bad debts are of greatest relative importance within machinery, the electrotechnical industry and iron and steel. Chemicals and rubber, glass, china and ceramics, construction materials and pulp and paper on the other hand belong to industries with a lower relative size of the bad debts burden. It is at the same time interesting to see that for some industries, e.g. metallurgy both ferrous and non-ferrous, pulp and paper, the printing industry or leather, furs and footwear, the bad debts are of secondary origin, i.e. the firms in these industries do not pay their suppliers (or generally creditors) because they do not get paid from their customers. In other industries, e.g. the electrotechnical industry but also glass, china and ceramics or clothing, a considerable share of bad debts are of primary origin. The variation coefficient calculated for the PBDS indicator at the same time shows that, for example, within the electrotechnical industry the relatively high value of this indicator is characteristic for most firms and on the other hand, the variability of this indicator is very high within non-ferrous metallurgy, though its average value is very small.

Looking at the bad debt indicators in different size categories of firms in table 4.5, it is perhaps rather surprising that the largest firms are those where the relative size of the primary bad debts is the smallest and where even the relative size of total bad debts as measured by the TBDS and TBDD ratio is smaller compared with firms in the second or third quartile. Perhaps some of the above speculations about the reasons for higher leverage of the big firms could explain this fact. They are perhaps more capable of repaying their maturing debts by newly extended credits by the banks. Thus, even though heavily indebted, they succeed in not being burdened by as many bad debts as smaller firms, which do not have such good access to new credits.

Table 4.4
Bad debt indicators for Czech industry in 1992

Industry	TBDS	PBDS	TBDD	TBDE	TBDA	PBDS var. coef.
Total	1.29	0.22	0.21	0.14	0.08	1
Iron and steel	1.67	0.06	0.31	0.17	0.11	2.36
Non-ferrous metallurgy	0.49	0.01	0.14	0.08	0.05	3.52
Chemicals and rubber	0.39	0.07	0.10	0.05	0.03	0.65
Engineering	1.99	0.42	0.24	0.21	0.09	0.75
Electrotechnical	2.31	0.68	0.27	0.21	0.11	0.49
Construction materials	0.41	0.12	0.11	0.02	0.02	1.66
Wood processing	0.74	0.11	0.19	0.10	0.07	0.98
Metal processing	0.95	0.12	0.21	0.13	0.08	0.52
Pulp and paper	0.46	0.01	0.09	0.05	0.03	2.22
Glass, china and ceramics	0.41	0.25	0.11	0.04	0.02	1.12
Textiles	1.15	0.22	0.22	0.13	0.08	0.38
Clothing	0.46	0.14	0.11	0.06	0.04	1.25
Leather, footwear and furs	0.99	0.04	0.20	0.12	0.08	0.86
Printing	0.58	0.02	0.14	0.08	0.05	2.54
Other industrial activities	0.98	0.07	0.22	0.12	0.08	0.34
Building and civil engineering	2.68	0.39	0.13	0.07	0.02	0.20

TBDS   total bad debts/monthly sales
PBDS   primary bad debts/monthly sales
TBDD   total bad debts/total debts
TBDE   total bad debts/equity
TBDA   total bad debts/assets
Source: Czech Ministry of Industry and Trade database, own calculations.

Table 4.5
Bad debt indicators in different size categories of firms in 1992

	Number of firms	TBDS	PBDS	TBDD	TBDE
Total	1,050	1.29	0.22	0.21	0.14
1st quartile	biggest 50	1.28	0.16	0.20	0.14
	212	1.23	0.27	0.21	0.14
2nd quartile	262	1.41	0.25	0.24	0.14
3rd quartile	262	1.46	0.25	0.25	0.14
4th quartile	264	1.16	0.30	0.23	0.10

Source: Czech Ministry of Industry and Trade database, own calculations.

We have mentioned above that high leverage of firms leads to their financial fragility, a phenomenon which might prove unfavorable especially in periods of great economic fluctuations and external shocks. The values of indebtedness of Czech firms in the above tables are not very different from similar indicators for enterprises in western countries. Some caution, however, is to be suggested. We have been using for our calculations as mentioned above the book values of the balance sheet variables. Bernanke and Campbell (1988), show that even in stabilized western economies the same indicators calculated on the basis of book and market values can be quite different. This is more than relevant for our situation. We are quite sure that the market values of the assets of the Czech enterprises are in most cases very different from their book values and thus the above calculated indicators do not necessarily reflect the true financial leverage of the firms and thus do not provide us with a complete picture about their financial stability.

Beside the impact of leverage on financial stability of enterprises and consequently on economic growth, another important linkage is between enterprise debts and inflation. Here we must distinguish two aspects. On the one hand an inflationary rise in prices, or more precisely, higher than expected inflation, diminishes the debt burden for the debtors and thus redistributes wealth from creditors to firms. This may positively affect economic growth as firms will be willing to extend investment and output and perhaps even the creditors, forgetting the above unfavorable wealth effect, will be willing to extend credits to higher net worth (and thus lower risk) businesses. This inflationary effect helped to diminish the size of the debt burden for the debtors in the world economy in the 1970s and it has helped to dwarf the size of debts in the neighboring high inflation of Poland or Hungary in the first phases of economic reforms. The success of the Czech Republic (or former Czechoslovakia) in fighting inflation did not help our debtors in this respect very much.

Table 4.6
Share of financial costs in total output in nonfinancial enterprises in 1991 and 1992 (%)

Industry	1991	1992
Total nonfinancial sector	13.33	13.39
Mining and quarrying	16.49	16.65
Food and tobacco	5.43	6.24
Textiles and apparel	10.56	11.89
Coal, oil and nuclear fuels	1.93	1.49
Chemicals	7.34	8.25
Metalworking	8.49	9.69
Machinery and instruments	15.05	17.16
Production of means of transportation	12.99	12.85
Gas and electricity	4.52	5.54
Construction	12.33	12.38
Transportation, warehousing and telecommunications	12.10	12.53

Source: Czech Statistical Office.

On the other hand, inflation raises the financial costs associated with debt due to the rise in interest rates (see Abel and Siklos, 1993). More precisely, the real interest rates being fixed by the banking sector, the percentage increase in nominal interest rates is much more dramatic compared to other nominal variables like income of the enterprises, wages, material costs etc. For example, when inflation is 20 per cent we can assume in a very simplistic case that when all prices accommodate to this inflation rate, that sales, wages or material costs of an enterprise will rise exactly by 20 per cent. The nominal interest payments, the real interest rate being held for example at 5 per cent, will on the other hand rise by 400 per cent from preinflation 5 per cent to the inflationary 25 per cent. The nominal sums of paid interest during periods of increased inflation thus represent a much more serious financial burden compared with periods of low inflation and of course, this burden is high especially for highly leveraged firms.

Table 4.7 illustrates the increase in financial costs of the enterprises in the 'inflationary' year 1991.

Table 4.7
Share of financial costs in total output in nonfinancial enterprises in 1991 and 1992 (%)

	Jan.-March	Jan.-June	Jan.-Sept.	Jan.-Dec.
1991	10.98	12.02	12.79	13.33
1992	12.62	13.16	13.00	13.39

Source: Czech Statistical Office.

Liquidity and profitability

The firm's insolvency preceded by the alarming development of the leverage ratios is the extreme case of financial problems of the company. The firms might on the other hand get into financial distress even when the leverage ratios are quite optimistic due to the uncoordinated cash flow and the problems of liquidity. The various liquidity ratios in table 4.8 illustrate the financial situation of Czech industry in this respect. The ratios calculated in the table are: CR, QR, CUR and PI.

The different liquidity ratios are more favorable within construction, construction materials or glass, china and ceramics and less favorable within metal processing, electrotechnical industry or textiles.

Within the size distribution of firms as given in table 4.9 the three liquidity ratios (CR, QR and CUR) are the lowest within the group of the second largest firms.

Table 4.8
Indicators of liquidity in Czech industry in 1992

Industry	CR	QR	CUR	PI	CR var. coef.
Total	0.19	1.08	1.50	1.97	1
Iron and steel	0.07	1.12	1.18	1.65	0.07
Non-ferrous metallurgy	0.12	1.24	1.35	2.80	0.07
Chemicals and rubber	0.15	0.95	1.11	8.84	0.22
Engineering	0.26	1.11	1.81	0.60	0.46
Electrotechnical	0.11	0.81	1.08	0.53	0.62
Construction materials	0.64	1.96	2.16	5.12	0.33
Wood processing	0.15	0.88	1.09	1.86	0.28
Metal processing	0.13	0.81	0.95	2.43	0.74
Pulp and paper	0.12	0.95	1.03	1.36	0.10
Glass, china and ceramics	0.36	1.47	1.93	4.49	0.26
Textiles	0.12	0.82	1.08	1.12	0.74
Clothing	0.20	0.98	1.39	2.22	0.12
Leather, footwear and furs	0.06	1.22	1.37	0.56	0.29
Printing	0.18	1.04	1.08	3.62	0.34
Other industrial activities	0.21	1.19	1.28	3.84	0.55
Building and civil engineer.	0.40	1.58	2.53	3.42	1.06

CR (cash/ratio)   cash/current liabilities
QR (quick/ratio)   cash + securities + receivables/current liabilities
CUR (current ratio)   cash + securities + receivables + inventories/current liabilities
PI   profits before taxes/interest payable
Source: Czech Ministry of Industry and Trade database, own calculations.

However, some caution should be expressed. Not only because the ratios are calculated using book values but as well because it is not quite clear how liquid in reality the assets included in these ratios are, namely receivables and inventories. It might be quite true that QR and CUR overestimate the real liquidity of the firms. On the other hand it is not clear whether high values of the liquidity ratios are desirable. Rather the question is about the optimal value of the indicators. Both too high and too low values might be the symptoms of some financial problems. The low values of low liquidity and the high values of poor financial management, when some of the assets could be used for more productive purposes, are problematic.

Quite interesting is that the indicator of interest coverage by profits (PI) is most favorable for the smallest firms and least favorable for the large ones. This is explained by the higher profitability of the smaller firms on the one hand and the lower share of bank credit in total debts in these firms on the other hand.

Table 4.9
Liquidity indicators in different size categories of firms in 1992

	Number of Firms	CR	QR	CUR	PI
Total	1,050	0.19	1.08	1.50	1.97
1st quartile	biggest 50	0.23	1.17	1.69	0.78
	212	0.15	0.97	1.33	1.87
2nd quartile	262	0.18	1.06	1.39	2.48
3rd quartile	262	0.25	1.15	1.58	2.15
4th quartile	264	0.25	1.22	1.50	2.67

Source: Czech Ministry of Industry and Trade database, own calculations.

Let us finally look at the profitability and activity indicators. The chemical, rubber and printing industries are those with high profitability and especially within the chemical and rubber industries high profitability is in accordance with the variation coefficient typical for all firms. On the other hand profitability within the engineering and electrotechnical industry, though on average low, differs considerably among different firms. Profitability is low as well within leather, furs and footwear. Among firms of different sizes profitability is lowest in the largest firms.

As in the case of the other ratios we should not overestimate the profitability indicators in their role of illustrating the viability of firms and industries. We have found in our interviews with managers that most firms at present give priority to business deals which promise guaranteed payments for the goods rather than profitable but possibly risky deals.

Table 4.10
Profitability and activity indicators for Czech industry in 1992

Industry	PE	PA	SA	PE var. coef.
Total	0.07	0.04	0.69	1
Iron and steel	0.06	0.03	0.75	0.19
Non-ferrous metallurgy	0.12	0.07	1.15	0.10
Chemicals and rubber	0.17	0.11	1.02	0.08
Engineering	0.03	0.01	0.57	0.61
Electrotechnical	0.03	0.01	0.59	1.02
Construction materials	0.08	0.06	0.58	0.25
Wood processing	0.07	0.05	1.07	0.47
Metal processing	0.11	0.07	0.98	0.12
Pulp and paper	0.04	0.03	0.83	0.39
Glass, china and ceramics	0.12	0.07	0.67	0.19
Textiles	0.05	0.03	0.83	0.42
Clothing	0.12	0.07	1.02	0.10
Leather, footwear and furs	0.03	0.02	0.88	1.61
Printing	0.13	0.08	1.02	0.24
Other industrial activities	0.15	0.10	0.94	0.20
Building and civil engineering	0.09	0.03	0.11	2.07

PE profits before taxes/equity
PA profits before taxes/total assets
SA sales/total assets
Source: Czech Ministry of Industry and Trade database, own calculations.

Table 4.11
Profitability and activity ratios in different size categories of firms in 1992

	Number of firms	PE	PA	SA
Total	1,050	0.07	0.04	0.69
1st quartile	biggest 50	0.06	0.03	0.68
	212	0.08	0.04	0.72
2nd quartile	262	0.08	0.05	0.65
3rd quartile	262	0.07	0.04	0.62
4th quartile	264	0.07	0.05	0.66

Source: Czech Ministry of Industry and Trade database, own calculations.

Banks

Role of banks

In the first part of this paper we discussed the importance of financial structure of the enterprises for corporate governance. The basic principle is that if somebody puts some money into a firm (into an investment project of a firm), he is interested in maximum stream of future profits, i.e. in maximum value of the firm. Depending on whether this financier comes from inside or outside of the firm and in what form the funding is provided, a concrete motivation framework for the management of the enterprise comes into being.

In our conditions the above logic has some specificities. Privatizing through voucher privatization means that the new owners put nothing into the privatized firms, which certainly affects their motivation to follow what happens with to their property, that is, to monitor the managers and make them maximize the value of the firm. It would be quite reasonable to expect that in many cases the new owners who do not risk their money will not be very motivated to monitor managerial efforts. For some of the institutional 'investors' with wide portfolios of owned corporate shares such monitoring will not be even feasible. On the other hand they might be interested in making the managers start even very risky and possibly very profitable (or on the contrary loss making) investment projects. The arising structure of ownership would thus lead to a motivation framework which would not lead to an optimum combination of managerial effort and risk taking.

For the same reasons many of the voucher owners may follow short-term strategies maximizing the stream of their short-term incomes, rather than longer term development strategies. On the other hand, because in the course of voucher privatization there is no direct inflow of funds into the firm, managers cannot appropriate any additional pecuniary or nonpecuniary benefits resulting from additional investments and thus the only sources of these benefits for them are the existing assets of the firm. The economic and financial situation of the enterprises might in this connection further deteriorate.

The above paragraphs do not sound very optimistic and if the outcome of voucher privatization were really the described incentive framework the voucher scheme would not be a big success in terms of future development of the privatized enterprises. Fortunately one major qualification has to be made. The new ownership pattern is not the only element shaping the motivation framework for the enterprises. The banking sector as the creditor of industry plays an important role as well and in our opinion its role in corporate governance will be further strengthened.

There are several reasons for this belief. First as indicated in the previous section the banking sector is the holder of more than half of the total enterprise debt. The banking sector has already in the past put money into the firms, a considerable amount of money, and thus is quite naturally motivated to follow what happens with the enterprises and even motivated to influence their behavior. In this connection, the fact that in many cases, the commercial banks or the investment companies they have created are the founders of the investment privatization funds, is a favorable development. The banking sector in this way combines the external ownership of enterprises with the debt contract which modifies the above described pessimistic outcome of the voucher scheme. We may even speculate that this combination might lead to an incentive framework for the management of enterprises which would integrate the positive features of external ownership and debt contract concerning the appropriate mix of effort and risk undertaken by the managers. This could be true especially in enterprises where the management would hold some stake in the company.

Second, the comparative advantage of the commercial banks compared with other big creditors which are the enterprises themselves is that the debt contracts with banks are concluded with professionals under standardized terms, where the rights and duties of the parties are clearly specified, whereas inter-enterprise debt often arises under more vague conditions. The monitoring activity of the commercial banks is an integral part of the debt contracts they conclude. In the interviews with the managers of several industrial enterprises we have found e.g. that a regular monthly check of the books of the companies is usual procedure included in current credit contracts. The screening and monitoring activity associated with the investment credit should be theoretically even more exhaustive. This puts the commercial banks into the position of strong principals capable of influencing very directly the behavior of the enterprises.

Third, the importance of the banking sector will become even more obvious in the postprivatization period. The privatized enterprises to which there was no inflow of financial resources during voucher privatization will be in need of considerable volumes of external finance to cover their restructuring and investment requirements. We are rather skeptical that the emerging capital market could provide the needed funding. Not perhaps due to the absolute shortage of funds in the market but because issuing stocks and bonds is not the way to obtain the needed external finance in general but usually only for established and well performing firms. In our situation we expect that tapping the capital market's financial resources in this way will be the privilege of only a limited number of companies. After all even in western market economies the capital market as provider of external finance for corporate growth is of secondary importance.

Let us now look at the role of the banking sector and its policies towards borrowers more closely. The hypothesis to start with is that the size of the financial sector in an economy, its level of development and its structure reflect the development stage of the economy. This is not surprising assuming that a developed and well structured financial system is a necessary precondition for economic growth and on the other hand dynamic economic development stimulates the growth of adequate financial intermediaries. King and Levine (1993, pp. 156-189), in an interesting empirical analysis have confirmed this, showing that the financial sector is far more important in developed countries compared with poor countries, that the weight of the central bank in credit activity is much smaller in these countries and that the financial sector in rich countries is more focused on private borrowers.

Similar results were achieved looking at the different countries in dynamic perspective. The large commercial banks sector and large shares of the private sector in total borrowing are the characteristics of rapidly growing countries.

We have attempted to calculate ratios similar to those of King and Levine (table 4.13). Though there might be some small methodological differences due to the lack of information on King and Levine's calculations, and thus the absolute values for all the ratios might not hold, the trends are important as well. The numbers for the Czech Republic clearly show the increase of the commercial banks' share in the credit market, the gradual decline of the central bank in this respect and the rapid increase of the share of private borrowers in total bank credit. All these trends are characteristic if we move from the group of poorer countries to richer countries or from the slowly growing to the more rapidly growing economies.

Table 4.12
Selected financial indicators of 116 countries and the Czech Republic

Country	Indicator	CBY	BY	PRIVY	BANK	PRIVATE
Very rich[a]	(29)	0.07	0.66	0.53	0.91	0.71
Rich	(29)	0.16	0.39	0.31	0.73	0.58
Poor	(29)	0.27	0.28	0.20	0.57	0.47
Very poor	(29)	0.17	0.19	0.13	0.52	0.37
Growth[b]						
very fast	(29)	0.11	0.46	0.35	0.81	0.70
fast	(28)	0.10	0.33	0.27	0.73	0.56
slow	(29)	0.10	0.24	0.2	0.71	0.61
very slow	(28)	0.12	0.17	0.13	0.60	0.51
Czech Republic						
Dec. 1991		0.37	0.69	0.11	0.65	0.10
Dec. 1992		0.37	0.74	0.23	0.67	0.21
June 1993[c]		0.22	0.75	0.26	0.77	0.26
Sep. 1993[d]		0.21	0.75	0.29	0.78	0.30

a   1985
b   1960-89
c   GDP = GDP (Q3 1992 + Q4 1992 + Q1 1993 + Q2 1993)
d   GDP = GDP (Q4 1992 + Q1 1993 + Q2 1993 + Q3 1993)
CBY   Central Bank Domestic Credit/GDP, for the Czech Republic: Central Bank Credit = Total Assets - Foreign assets - Cash within the Central Bank balance sheet
BY   Commercial Banks Domestic Credit/GDP
PRIVY   Gross   Credit   to   the   Nonfinancial   Private   Sector/GDP
BANK   Commercial   Banks   Domestic   Credit/Commercial   Banks   + Central Bank Domestic Credit
PRIVATE Gross Credit to the Nonfinancial Private Sector/Total Domestic Credit (incl. Central Bank domestic credit)
Source: King, R. G. and Levine, R. (1993). Own calculations using Czech National Bank and Czech Statistical Office data.

The above ratios on the other hand do not say very much about the quality of assets held by the banks, about the relation between demand and supply of funds in the credit markets or the quality and diversity of services provided by the banking sector. In this respect the Czech banking sector is rather underdeveloped and faces several difficulties.

The big problem is the quality of assets held by the commercial banks. The poor quality of many of the credit items in the balances of the banks is not only due to the old bad loans provided under the system of central planning. In fact the share of old bad loans in total bad loans permanently decreases.

Table 4.13
Percentage shares of risk credits[*] in total credits of the commercial banks

December 1991	2.4
June 1992	3.9
December 1992	190.
June 1993	21.7
October 1993	23.1
of which:	
Short-term credit	31.6
Medium-term credit	13.6
Long-term credit	19.2

[*]   risk credits = temporarily illiquid claims + dubious credits
Source: Own calculations using Czech National Bank data.

In many cases the bad loans came into being recently due to the unpredictable developments in the enterprise sector in the transition period but also due to the poor management of the credit portfolios by the banks. This is often the case of the new small banks which sometimes boast of their dynamic credit activity and in fact accumulate in their balances many dubious assets.

Starting from 1991 banks evaluate the liquidity of their assets on the basis of unified methodology. A special focus is on risk credits, comprising several categories of loans which either became temporarily illiquid or are already nonperforming. The results are summarized in table 4.13. They indicate that at the end of 1993 nearly one quarter of the stock of credits granted by banks were classified as risk credits. The sharp increase in this share since 1991 does not mean that the structure of assets of the banks has deteriorated so dramatically. It only means that the banks in the past have not included in their data many of the already existing bad loans.

The biggest share of risk credits is within the category of short-term credits. This is not to say that short-term credits are the most risky credits or that they are provided with the least care. It is only a natural result of the fact that short-term bad loans fall into the category of illiquid or nonperforming loans much earlier compared with medium-term or long-term credits.

The commercial banks facing the above problems of bad loans have adopted several strategies. One is the preference given to short-term as opposed to long-term credits. This is reflected in the changing structure of the stock of bank credits as illustrated in table 4.14. The short-term credits are believed to be of lower risk as it is assumed that the conditions of a viable firm cannot change dramatically in the short-term. In the longer term, however, the ability of the firm to meet its financial obligations is far less certain. At the same time short-term credits are the way in which the banks start their credit activity vis-à-vis new clients. Once they verify their financial viability they may extend their credit activity towards longer term loans. On the other hand the above restructuring of bank assets reflects the changing composition of the banks' liabilities. The banks are reluctant to extend long-term credits due to the lack of long-term financial resources.

Table 4.14
Structure of the bank credits according to their maturity (%)

Type of credits	6/91	12/91	6/92	12/93	6/93	10/93
Short-term	36.4	39.5	40.3	37.4	41.8	44.1
Medium-term	17.4	18.2	20.9	26.7	28.2	27.2
Long-term	46.2	42.3	38.8	35.9	30.0	28.7

Source: Own calculations using Czech National Bank data.

The restructuring of the credit market is even more obvious if we focus on some specific categories of credit, e.g. investment credit.

Table 4.15
Investment credit by maturity between December 1991 and October 1993 (in billion CK)

	12/91	6/92	12/92	6/93	10/93
Short-term	1.7	4.7	9.0	14.1	16.4
Medium-term	36.5	44.2	56.0	65.9	71.0
Long-term	71.8	71.3	63.0	63.7	71.2

Source: Czech National Bank.

Between December 1991 and October 1993 the short-term investment credit in nominal terms increased nearly ten times, the medium-term investment credit doubled and the long-term investment credit stagnated.

The banks' accommodation to the bad loans phenomenon was furthermore in terms of their interest rates policy. One typical feature of this policy was the considerable differences between the interest rates on credits and deposits.

Though declining in relative terms (as a per cent of interest rates on deposits) since the third quarter of 1992, the spreads between credit and deposit rates still remain very high.

The other typical feature of the commercial banks interest rate policy is their differentiated approach towards different borrowers. This differentiated policy is applied in the course of dramatic restructuring of the credit market as concerns the ownership structure of the borrowing community.

The share of private businesses in total credit provided by the commercial banks increased from 16.4 per cent in June 1991 to 46.8 per cent in October 1993 and the share of state-owned borrowers declined from 76.0 to 39.0 per cent. In some cases the new private clients are the old firms which were privatized. The old firms, however, in most cases underwent some organizational restructuring, some companies were divided into several new ones and in most cases the management changed. We can thus argue that in general the banking sector faces, within the private sector, new clients who do not have a reliable past financial record and which have to build their links with the banks anew.

Table 4.16
Spreads between average interest rates on total credits and total deposits

	3Q/92	4Q/92	1Q/93	2Q/93	3Q/93
Percentage points	7.31	6.91	7.22	7.48	7.09
As per cent of interest rate on deposits	120.40	109.70	104.60	103.00	101.40

Source: Own calculations using Czech National Bank data.

Table 4.17
Structure of the credit market regarding the ownership type of the borrowers (%)

	6/91	12/91	6/92	12/92	6/93	10/93
State owned	76.0	69.1	58.3	49.6	44.7	39.0
Private incl. co-ops	16.4	22.3	27.6	36.8	40.7	46.8
Foreign control. enterpr.	0.0	0.1	2.0	2.6	2.5	2.9
Other*	7.6	8.8	12.1	11.1	12.2	11.3

* population, credits abroad, other nonclassified credits
Source: Own calculations based on Czech National Bank data.

In this situation it is not very surprising that the banks are still willing to credit the old large enterprises and that they are rather cautious when granting loans to the new private firms. This is valid more for the old large banks with established ties with the SOEs than for the new private banks for whom the private firms are often the only clients and which sometimes are very uncritical in providing these firms with credits.

For the big banks it is also easier to manage a few large credit items to 'well known' old clients than a large portfolio of minor loans to many smaller and so far unknown borrowers. Last but not least the motivation of the big banks to credit the old large firms comes from the fact that the banks are the holders of considerable past debts in these firms and thus not providing them with new credits could endanger the liquidity of these assets in the balances of the banks.

Table 4.18
Interest rates on credits to state-owned and private borrowers

	State-owned	Private incl. cooperative
3rd quartile 1992	13.62	15.09
4th quartile 1992	13.18	15.18
1st quartile 1993	14.88	14.73
2nd quartile 1993	15.65	15.16
3rd quartile 1993	14.46	14.64

Source: Czech National Bank.

This differentiated approach was evident especially at the beginning of the privatization process. Later on as some of the old SOEs became privatized and the private firms stabilized, the interest rates on credits to the different ownership type borrowers move together and, in the beginning of 1993, it was already the private sector which in terms of average interest rates on credits was treated more favorably by the banking sector.

The adjustment of the bank's policies towards the borrowers also has the form of more strict conditions of the credit contract. We have so far only scattered information in this respect based on our survey of a sample of firms but we expect that such conditions of credits as careful collateralization or detailed analysis of the company's books have become the standard.

We have mentioned above that the interest rates on credits seem to us quite high taking into account the level of interest rates on deposits. On the other hand there are some indications from the banking sphere that the demand for loans in spite of the given interest rates is still very high. In this respect perhaps the interest rates on credits are still quite low and some estimates are that the banks could allocate the existing amount of credit for interest rates even considerably higher than those existing. The banks do not do this, however, and we are not quite sure what their reasons are for this. Perhaps they are constrained by the Central Bank in their interest rate policy or they simply assume it not to be a good policy to raise the interest rates now, when it is expected that in the near future they should decline.[4]

Conclusion

The fact that within voucher privatization the banks emerge both as creditors and holders of considerable shares of stock of the privatized enterprises has one important implication. It favors the rise of a bank based financial system rather than a capital market based system of supply of external finance to enterprises.

The reason is simply that the banks, being the creditors of the firms, will be interested in affecting their behavior by means of more stable, longer-term ownership control. The voucher privatization thus, rather than leading to the emergence of a financing and monitoring mechanism by impersonal capital markets, will lead to more stable and tight interlinkage of the banking sector with enterprises.

What is the relevance of this fact from the point of view of the theoretical framework discussed earlier? In the first part we have distinguished between external finance by means of debt and equity as two different arrangements of financing enterprises from external sources with different incentive outcomes as concerns the behavior of the enterprise management. The first thing which should be stressed in the above context is that the dividing line between debt and equity as two components in the financial structure of the firm does not represent the dividing line between the banking sector and the owners of the firms which acquire their property rights via capital markets (or via the voucher scheme). On the one hand, as it is becoming the reality in this country and as it is the case in other developed countries as well, e.g. Japan or Germany, the banks fulfill both the roles of creditors of the companies and their stockholders. On the other hand debt may have the form of credit provided by the bank or the form of bonds sold in the capital market (moreover also the form of trade credit). So perhaps in reality it is not as important what form external finance acquires as by whom it is provided.

The second point is that the banks as financial intermediaries reduce the various costs related to financing from external sources. This fact explains why financial intermediaries actually exist. According to one approach which we have mentioned in the first section, the role of intermediaries is to diversify risk which otherwise would be directly borne by individual savers-investors. The advantage of the banks as financial intermediaries is that in their combined role of creditors and owners and due to their detailed information on the firms' activities they are able to perform this task highly efficiently. On the other hand it is true that due to the long-term credit relation to a specific group of firms banks might have an inclination to stream the financial resources to this group of firms which need not necessarily be the most efficient investments.

The second approach stresses the role of financial intermediaries in reducing agency costs due to more efficient monitoring. This is Diamond's explanation of financial intermediaries as institutions performing delegated monitoring. This fact stresses the principal-agent dimension of the relationship between the financial institution and the firm. The relation between the bank and the company however does not run in one direction. Rather in an ideal case it is a long-term partnership where information flows in both directions and where there are benefits for both partners.

Hellwig (1991), in this connection talks about financial intermediation as an instrument for the emergence of a binding relation between the banks and the firms. The commitment of both the banks and the enterprise to a long-term exclusive relationship is based on economic rationale. We have explained above that concentration of external finance reduces monitoring costs and thus the firm optimally has one main financier. As the relationship between the firm and the bank develops in the longer term both parties acquire information advantages compared to outsiders. For the bank it is less costly to provide financing to the already well known firm than to an unknown entity where costly information gathering would have to be implemented. The firm, on the other hand, seeing the commitment of the bank and its willingness to supply funds and even start rescue operations is not motivated to search for alternative financiers. The competition for the bank thus diminishes.

This decline in competition which occurs as a result of the commitment arrangement between the banks and the enterprises on the other hand represents the disadvantage of this stable relationship. The costs of this are in terms of less efficient allocation of funds by banks or less favorable loans for the firms. What matters then is the comparison of costs of the competitive and commitment arrangements. This rationale explains well why the large established companies with well known past economic and financial records often access the capital markets for external funding, whereas the less known firms obtain external finance predominantly in the form of bank credits.

The other problem of the commitment arrangement is that within the common financial strategy which the bank helps to shape the firm might feel that it loses independence and therefore it might want to break or limit these close links at some moment. This is why potential competition of other banks still plays its role. In this sense the market for commercial banks services is to some extent contestable.

Taking into account the above, what can we expect in the development of the Czech financial system in the short and medium run? We have already shown that privatization in the form it was implemented has strengthened the role of the banks within the financial system. We should not be misled in this connection by the dramatic development of the capital markets which quite recently did not exist at all.

The role of the banks will be further strengthened by the increased demand for external finance in the postprivatization restructuring period. There is no other source of finance which could meet this demand except for a few firms which will be able to tap the capital market and the 'international' firms which may exploit foreign credits. For the bulk of the economy domestic banks will play a dominant role.

We may expect furthermore some concentration in the stock portfolios of the IPFs which will further intensify the relationships between the banks as the founders of the IPFs and the enterprises.

Our interviews in enterprises have so far confirmed this, showing that the situation: one company - one bank generally characterizes the relations between the banking sector and industry. Attempts to exploit the competition among the banks by existing established firms are marginal.

In the longer term there will be some developments within the industrial sphere as well as the banking sector which might on the contrary lead to some loosening of the established links between the banks and the firms. One is the future ability of firms to finance their growth predominantly from internal sources which will lead quite naturally to some emancipation of enterprises from their relationship with the banks. On the other hand convertibility and further liberalization of capital flows leading to more competition by foreign banks might result in some erosion of the traditional bank-firm links as well. At least it will lead to more aggressive strategies of the banks in the credit markets.

Notes

1. The views expressed in this paper are those of the authors and do not represent those of the Czech National Bank
2. This is in contrast to some previous approaches. For example, according to the Modigliani-Miller theorem the structure of firm finance is irrelevant under assumptions of perfectly competitive capital markets, equal tax treatment of different sources of finance and symmetric information of economic agents. In such a world we could observe very different financial structures of firms having

no impact on their performance and their value. The assumptions are however quite unrealistic and in the real world we can observe certain similarities of financial structure of the firms or at least great attention paid by both the firms and their financiers to the choice of relevant financing methods. Modigliani and Miller, realizing this, have later argued that preferential tax treatment of financing by debt would increase the value of debt financed firms. Modigliani and Miller (1958) and (1963).

3   The maximizing assumption about the owner-manager may seem too strong as he may in fact, due to his position, choose any mix of effort and leisure, which of course happens, and affects the value of the firm, i.e. the owner-manager surely maximizes his utility function but not necessarily the value of the firm. The important thing is in any case that no additional agency costs arise in this respect.

4.   There are several factors which will put downward pressure on the interest rates in the near future, such as lower inflation, competition among the banks, liberalization of international capital transactions which will lead to more competition from abroad, increases in the banks' own capital as well as the policies of the Central Bank which has expressed its dissatisfaction with the level of interest rates.

References

Abel, I. and Siklos, P. L. (1993), 'Constraints on enterprise liquidity and their impact on the monetary sector in formerly centrally planned economies', *CEPR Discussion Paper*, no. 841.

Bernanke, B. S. and Campbell, J. Y. (1988), 'Is there a corporate debt crisis?' *Brookings Papers on Economic Activity*, no. 1, pp. 83-125.

Bernanke, B. S. and Gertler, M. (1990), 'Financial fragility and economic performance', *The Quarterly Journal of Economics*, February, pp. 87-114.

Czechoslovak Statistical Office (1992), *Financing of Investment, Focus on Public Investment* (in Czech), Prague.

Davis, E. P. (1992), *Debt, financial fragility and systemic risk*, Clarendon Press, Oxford.

Diamond, D. W. (1984), 'Financial intermediation and delegated monitoring', *Review of Economic Studies*, L1, pp. 393-414.

Edwards, J. (1993), *The financing of industry, 1970-89: an international comparison*, unpublished manuscript.

Gerschenkron, A. (1962), *Economic backwardness in historical perspective*, Harvard University Press, Cambridge, MA.

Grossman, S. J. and Hart, O. D. (1980), 'Takeover bids, the free-rider problem, and the theory of the corporation', *Bell Journal of Economics*, no. 11, pp. 42-69.

Hájek et al. (1993), 'The macroeconomic analysis of the Czech economy (Situation in the first half of 1993)(in Czech)' *Working Paper*, no. 14, Czech National Bank, Institute of Economics.

Hellwig, M. (1991), 'Banking, financial intermediation and corporate finance' in Giovannini, A. and Mayer, C. (eds), *European financial integration*, Cambridge University Press, Cambridge.

Jensen, M. C. and Meckling, W. H. (1976), 'Theory of the firm: managerial behavior, agency costs and ownership structure', *Journal of Financial Economics*, vol. 3.

King, R. G. and Levine R. (1993), 'Financial intermediation and financial development', in Mayer, C. and Vives, X. (eds), *Capital markets and financial intermediation*, Cambridge University Press, Cambridge.

Mayhew, K. and Seabright, P. (1992), 'Incentives and the management of enterprises in economic transition: capital markets are not enough', *CEPR Discussion Paper*, no. 640.

Modigliani, F. and Miller, M. H. (1958), 'The cost of capital corporation finance and the theory of investment', *American Economic Review*, vol. 48.

Modigliani, F. and Miller, M. H. (1963), 'Corporate income taxes and the cost of capital: a correction', *American Economic Review*, vol. 53.

Phelps, E. S., Frydman, R., Rapaczynski, A., Shleifer, A. (1993), 'Needed mechanisms of corporate governance and finance in Eastern Europe', *Working Paper*, no. 1, European Bank for Reconstruction and Development, London.

Sabel, Ch. and Piore, M. (1984), *The second industrial divide*, Basic Books, New York.

Schumpeter, J. A. (1955), *The theory of economic development*, Harvard University Press, Cambridge, MA.

Stiglitz, J. E. and Weiss, A. (1981), 'Credit rationing in markets with imperfect information', *American Economic Review*, vol. 71, no. 3, pp. 393-410.

United States Department of Commerce, *Quarterly Financial Report for Manufacturing, Mining and Trade Corporations*, Third Quarter, 1986, cited from Brealey, R.A., Myers, S.C. (1988), *Principles of corporate finance*, 3rd ed., McGraw-Hill, New York.

5 State-owned enterprises in transition: prospects amidst crisis

Jan Mujzel

Introduction

In the course of the systemic transformation, the performance of state-owned enterprises has so far remained an important determinant of the economic situation and the degree to which the economy satisfies the people's needs. This will continue for at least a couple more years. At the end of 1993, there operated in Poland 5,924 state-owned enterprises, 958 commercial-code companies wholly or more than 50 per cent owned by the State Treasury, and 830 partnerships more than 50 per cent owned by state-controlled legal entities. The number of employees in the public sector totalled 6.1 million, which accounted for about 40 per cent of the aggregate employment in all branches of the economy, excluding the 'grey' sector, or 53.8 per cent of the employment outside the 'grey' sector and individual farming. In 1993, the state-owned sector generated, disregarding the 'grey' sector again, nearly 50 per cent of the GDP. The sales volume of the public sector amounted to 62.6 per cent of the total in the area of industry, 14.2 per cent in professional construction services (construction and assembly enterprises), and 55.7 per cent in transportation services. It remains the main source of revenue for the state budget, and the crisis that has hit a large proportion of state enterprises is most probably one of the main reasons behind the critical state of public finance in Poland.

Privatization reduces the dependence of the entire economy, and the budget in particular, on the public sector. This process may and should intensify, but its economic and logistic complexity, together with its political and social ramifications, slow it down considerably. J. Olszewski's government planned to privatize about one half of the assets of state-owned enterprises by the end of 1994. That was already one year behind the schedule drawn by the previous two governments, and even so the task seems today very difficult, if not downright impossible, to achieve.

117

But even if it were somehow carried out, the efficiency of state-owned enterprises, or the improvement of it would remain a key issue.

The chapter is organized as follows: section one describes the attempts that were made before transition to make the state-owned enterprises more efficient. Section two comments on the situation of the state-owned enterprises at the eve of transition. Section three explains the assumptions of the Polish transformation program about state-owned enterprise behavior. The main section of the paper, section four, is devoted to the analysis and identification of the changes in internal functioning and in the external environment that were important determinants of the behavior of state-owned enterprises. Sections five and six analyze respectively financial performance and management behavior. The conclusion is drawn in section seven.

Failed reforms

Since the mid 1950s, the communist parties in practically all countries of real socialism tried to introduce market-oriented economic reforms, one of the main goals of which was to effect systemic modifications that would render state-owned enterprises more efficient and acceptable to the people. The scope, intensity, and timing of these efforts varied, but regardless of all the differences, they had one thing in common: until the end of the 1980s they went on under the conditions of unquestionable dominance of the principles of real socialism: preponderance of state ownership, central planning as the fundamental mechanism of coordination, and negligible role of capital in the distribution of income. Another trait common to all those reforms was their lack of success, as they all failed to bring about an improvement of efficiency on the assumed and desired scale. No doubt this was first of all due to their half-hearted character and lack of consistency. But another important reason was the policy of the communist regimes, which would try to undo any more profound changes and restore the classical mechanisms of the command economy, as soon as political tensions were eased.

And yet each of those reforms left some lasting, positive marks: the system evolved towards a free-market, multi-sector model, although that evolution followed a zigzag path, and sometimes actually proceeded backwards. The economic system (including the mode of operation of enterprises) differed considerably in mid 1989 in Poland, Hungary, Yugoslavia, and most other countries of the socialist block, from the thoroughly totalitarian model of real socialism that characterized the first half of the 1950s.

It should be mentioned at the end of these introductory remarks that during the period in question in the countries of real socialism, there did not evolve any extreme forms of market-oriented transformations, aiming for 'fully democratic, free-market socialism'. It was generally assumed that

118

any transformation of that type would clear the way for gradual erosion of the principles of the entire system and allow an evolutionary transition towards a free-market, capitalist economy, marked, however, by sensitivity to social issues and a human face. In Poland, an outline of such a conception was included in the socioeconomic part of the Round Table Agreements. In mid 1989, in view of the breakthrough in political developments, the conception lost its practical relevance and remained a 'paper program'.

The starting point

Thus the government of Mazowiecki and Balcerowicz 'inherited' an economy dominated by state-owned enterprises, of a somewhat hybrid type. Due to the earlier reforms, they had acquired some market-type attributes, but both their internal rules of operation and their environment were predominantly a product of the command economy, state-controlled and permeated with politics. The market elements comprised first of all a greater freedom of decision allowed to enterprises, more objective financial regulations, and the possibility to establish commercial-code companies, also ones involving foreign capital. The tradition of the command economy, on the other hand, was represented, internally, by the workers' self-management (formed in 1981 under the very peculiar circumstances of that time), whose relationships with the professional management stirred up much controversy; politicization and paternalism of the state institutions of control (especially the so-called 'founding bodies') and financial authorities; lack of an adequate incentive system (especially a long-term one) for employees and management, based on the criterion of efficiency. This was combined with such external factors as weakness of the private sector, a wide range of official price controls, large deficits of supply, mainly inconvertible currency, administrative rationing of a whole array of material and financial resources, monopolistic relationships on the domestic market, administrative regulation of crucially important trade with Comecon countries, and gross underdevelopment of the financial infrastructure, manifesting itself, among other things, through the enormous extent of the state's control over highly monopolistic so-called commercial banks and their heavy dependence on politics.

As a natural consequence of such a state of affairs, both the efficiency and the competitiveness of state-owned enterprises were low, by international standards. The enterprises were very often left with outdated and depreciated production assets; cost-effectiveness was low on their scale of priorities; overemployment was widespread; the products, especially more highly processed ones, in most cases clearly fell short of world standards. Thus a comprehensive, radical restructuring of the ownership relations of state-owned enterprises, as well as their rules of operation and both technological and financial conditions, inevitably became one of the most pressing demands of the economic transformation.

Assumptions of the program

The starting point for the famous economic program formulated by the government in October 1989, the so-called Balcerowicz program, was the assumption that it is necessary to establish a market system similar to that existing in highly developed countries. This fundamental change of the economic system was to be effected, among other things, through an ownership transformation that ought to bring the ownership structure into line with the model functioning in developed countries and through extending the range of autonomy of state-owned enterprises.[1] The first letter of intent to the International Monetary Fund, drawn up by the Minister of Finance and President of the National Bank of Poland, provided a little more detail. It said, namely, that: "The main envisaged forms of ownership comprise joint-stock companies, employee-owned enterprises, and publicly-owned enterprises".[2] This sounds like a paraphrase of some conception of a mixed economy, strongly emphasizing both privatization and rationalization of state-owned firms.

Such a formulation of the general strategy of systemic transformation was surrounded from the very outset by a peculiar 'air of doctrine'. It can be seen in a number of documents, but first of all in the statements and views of some leading politicians involved in the economic reform. The heart of the matter was, presumably, the deeply rooted skepticism about the possibility of state-owned enterprises ever achieving a level of efficiency comparable to that of private ones, even if they were to undergo some radical reforms. Such an assumption was suggested, apart from various theoretical considerations, by the long series of failed reforms under real socialism and the discouraging performance of state-owned firms in many non-socialist countries, both developed and developing. In addition to this, there was a conviction that the already meagre efficiency of state-owned enterprises could deteriorate even further if, as was the case in Poland, employee representations were allowed to have a say in the decision-making process.

Such considerations gave rise to certain political orientations. It was believed, firstly, that the changes in the performance of state-owned enterprises necessitated by the efficiency requirement could be effected, to some extent, not so much by influencing the internal mechanisms operative in the enterprises and their governance system, as through a radical transformation of their environment. Secondly, it would be desirable, however, to eliminate from the internal structure of the enterprises the participatory type of management, represented by the workers' councils established in 1981. Thirdly, in view of the insurmountable limits on the efficiency of state-owned enterprises, it would be necessary to accelerate their privatization using all ways and means available, including free, or partly free, distribution of public assets. Any resulting drop in budget revenues or other social costs were dismissed as temporary phenomena, which would be more than adequately compensated for by the increased efficiency and expansion in the long run anyway.

It is to be expected that such assumptions determined to a great extent the policy on state-owned enterprises during the first stage of the transformation. Later on, there followed some change of attitude and different tendencies began to emerge in the policy of the authorities. This subject receives fuller treatment in the second part of the paper.

The years 1990-93

It would be erroneous to say that no important efficiency-stimulating adjustments were made in the internal and external arrangements determining the functioning of state-owned enterprises during the years 1990-93. Such adjustments no doubt did take place, and in the case of the external circumstances, they constituted a real breakthrough. The following subsections will be devoted to the identification and discussion of the main points of those changes, which were to become one of the main determinants of the situation and fate of state-owned enterprises.

Organizational principles of the economic system

Apart from privatization, which is discussed separately, two other significant and in essence beneficial organizational changes occurred in the economic system. These were: rapid development of commercial-code companies, and the introduction of so-called managerial contracts.

The joint-stock company and limited liability company as the forms of organization, ownership, and types of legal status dominating large and medium-size enterprises in capitalism have long proved to have their important advantages. They allow flexibility of organization and offer attractive opportunities for accumulating capital and combining it with knowledge and managerial skills. Besides, in the restructuring of the public sector, they make it possible to eliminate the outdated arrangements based on workers' self-management, and open up many different paths of privatization. In the past three years there have emerged in this sector both relatively numerous Treasury-owned commercial-code companies, which are officially identified with commercialization of state-owned enterprises and are, formally, their interim form en route to full privatization, and more dynamic, mixed-capital companies, owned partly by the state and partly by domestic and foreign private investors. Until the end of 1993, 522 enterprises had been transformed into Treasury-owned commercial-code companies. Towards the end of the same year, the Treasury had a majority share in 958 companies, while various state-controlled legal persons - mostly state-owned enterprises - participated as majority share-holders in 830 companies.

Managerial contracts are meant to be concluded between groups of professional managers and the state as the owner. They should typically remain in force for a couple of years, during which the managers would usually also be supposed to restructure the enterprise. Such arrangements promise to stimulate efficiency in two ways: by eliminating the traditional division of competence between professional management and workers' councils, and by creating a true efficiency-based system of incentives not only for the managers, but also for the employees, as well as the boards of directors. In spite of the existence of suitable legal provisions since mid 1991, managerial contracts remained until 1992 at the stage of discussion and planning. In the first months of 1993, five Treasury-owned commercial-code companies were introducing this system, while a dozen more were getting prepared to do so. Until now the situation has not changed substantially. Presumably, the introduction of managerial contracts to non-transformed enterprises is being done on an even smaller scale.

It is not as yet possible to reliably assess the impact the transformation of state-owned enterprises into stock companies has had on efficiency, due to the short period in which to gather data and the shortage of serious comparative studies. One can expect that the consequences of the change should be, in general, favorable, but their scale is limited by the transitory status of a majority of the companies in question (the Treasury-owned commercial-code companies) and the insufficiency of complementary changes in other areas, particularly that of finance and Treasury's institutional framework.[3] Furthermore, the presumably great potential of managerial contracts for boosting efficiency has hardly been exploited in practice so far.

Privatization

Both theoretical reflection and experience seem to indicate that the influence of privatization on the performance of state-owned enterprises is multi-faceted and multi-directional, but all in all usually improves efficiency. It also appears that this influence and its final effects depend very heavily on the form of privatization and the manner in which it was carried out.

Let us take a closer look at this problem. In the quest for efficiency, one can notice some general tendencies present, typically, in any privatization anywhere, and some specific ones, which follow from the privatization procedures peculiar to Poland. The general tendencies include first of all the contribution of privatization to the creation of an efficiency-stimulating environment for state-owned enterprises. This point is further developed later on in this chapter. There is also the positive feedback linking the standing of a state-owned enterprise, its privatization, and the impact of privatization on efficiency, including the promise of higher wages and new jobs. Under the Polish conditions both types of influence occur, but the latter is weakened by the slowness of capital privatization (privatization through sales of equity), the principal form of privatization of large enterprises, and the lukewarm interest of foreign investors, whose involvement would bring about a form of privatization particularly desirable from the point of view of efficiency.

On the other hand, there are some positive, specifically Polish trends, such as changes in the organization of state-owned enterprises, transformation of traditional enterprises into Treasury-owned commercial-code companies, and managerial contracts, which are a case of privatization combined with structural transformation. The matter has been discussed in the previous section, along with the circumstances in 1990-93 that limited potential of both these factors to encourage greater efficiency.

Another important, in my opinion, and advantageous arrangement that is specifically Polish is the regulation, accepted after negotiation in 1990, distributing the prerogative to take decisions on privatization. It grants the director and the workers' self-management of an enterprise offered for privatization the right to take the initiative to privatize or, conversely, to oppose any such move. This concerns enterprises which are to be converted into Treasury-owned companies or privatized through liquidation.[4] Legally, the authorities have the power to override or get around the decisions of the enterprise's official representatives, but until recently they tried not to make use of it. However, since mid 1991, the ideas of acceleration of ownership transformations, mass privatization, curtailment of leasing by employees as the dominant form of privatization, and some other circumstances that will be discussed below, have produced a tendency to develop and embrace a more centralized model of privatization.

Participation of the managers and employees in the decision about embarking on privatization may greatly improve efficiency of both the enterprise in question and the privatization process itself. It is likely to introduce a sense of stability into the operation of the enterprise, eliminate sources of conflict, and hence improve efficiency. It can also contribute to the privatization process - through new, useful ideas, initiatives, and massive involvement. In 1990-93, this potential for efficiency stimulation was quite strongly suppressed and, consequently, largely wasted. The reason was the failure to effect a radical reform of the internal operating principles of state-owned enterprises which might prepare them, as far as possible, to meet the efficiency requirement under the conditions of transformation. This could have had, or, even, must have had an adverse effect on the decisions and performance of enterprises. On the part of the authorities, in turn, such a situation breeds a tendency to limit the rights of the enterprise and employees, and to resort to the 'carrot-and-stick' policy to goad them into privatization.

A third specific source of efficiency potential is probably to be found in a system of preferences and discounts for employees and managers who acquire property rights to their privatized enterprises. This is most clearly demonstrated by the case of capital privatization of state-owned enterprises, whose employees have so far enjoyed the right to buy up to 20 per cent of the stock at half price. The efficiency potential may also increase when instead of the reduction of share prices, some relatively smaller proportion of the shares is transmitted to the employees free of charge. (In accordance with the Common Privatization Program,[5] it will be up to 15 per cent.) The impact of both arrangements outlined above is reduced by the otherwise necessary legal ceiling on the maximum aggregate gain achieved in that way, equal to the average wage multiplied by the number of employees. It is doubtful, on the other hand, whether efficiency can be stimulated by an alternative form of privatization and preferences for the employees: the possibility of the enterprise or some part of it being leased for a couple of years on preferential terms to an employee-partnership, in which the managers usually have a larger share[6] and some outside investors can participate as well.

Privatization can also have a negative influence on the performance of existing state-owned enterprises. One of the mechanisms of this is connected with the feeling of mounting uncertainty, which blocks efforts to increase efficiency, especially long-range ones. It appears, however, that if such phenomena did occur in Poland, it was mostly under extreme circumstances only, like the prolonged political negotiations and controversies over the Common Privatization Program, the Pact on the State-Owned Enterprise, or changes in the basic privatization law. Another negative mechanism might involve forms and procedures of privatization that make the employees and managers lose interest in improving the efficiency of their enterprise, or even, temporarily, may wish that its performance would actually decline. In the years 1990-93 in Poland, no anti-efficiency phenomena of such an extreme type occurred on any significant scale.

Finally, it must be stressed that the conclusions and opinions presented in this section are mostly hypotheses. The empirical research carried out so far does not offer reliable verification of these hypotheses, one of the reasons of which is that separating the influence of privatization from many other determinants affecting the performance of state-owned enterprises can be extremely difficult, if not impossible. Nevertheless, in view of what has been said so far, the following conclusion seems to be quite well substantiated: privatization may, and usually does make the economic behavior of existing state-owned enterprises more rational and improves their efficiency. In the past four years of transformation in Poland, the opportunities offered by privatization have not been fully exploited, but even so, the beneficial influence of privatization has occurred and has been fairly significant.

Owners' institutions of supervision

In 1990-93, supervision over state-owned enterprises was exercised mainly by the traditional founding bodies, established during the abortive reform of the 1980s. Such institutions, branch ministries and province-level offices of state administration, stuck to the logic of the old system and turned out to be inadequate for the needs of the economy in transition. The multiplicity of decision-making centers affected the consistency of the economic policy, which is often a precondition of efficiency. It was quite natural for these institutions to be strongly involved in politics, and their functions, both as owners and regulators, had been tailored accordingly. This resulted in the traditional paternalistic attitudes and maintained monopolistic tendencies.

Things looked hardly better in the case of the other main supervisory institution - the Ministry of Ownership Transformations - which exercised control over the part of public property that had already undergone transformation. With the growing number of Treasury-owned stock companies and ones where the Treasury held minority interest, the Ministry turned out to be inadequate in its present form to the role of a modern-type owner of a large, quickly growing and highly differentiated sector of the economy. It has been understaffed and concentrated on the liquidation of state-owned property. The Ministry's inadequacy was reflected, for instance, in the weakness of the boards of directors of state-owned companies it appointed.

Thus the two main institutions of owners' supervision proved to be a weak link in the privatization process and an inadequate instrument in exerting efficiency pressure on state-owned enterprises, including the elimination of those of their number which had no chance to become competitive and survive on their own. This view is corroborated by numerous analyses and studies. A more optimistic development in this field was the establishment of the Agency for Agricultural Property of the Treasury, which may have marked the beginning of welcome changes in the structure of public institutions of ownership. Its first results seem encouraging.[7]

Labor relations

The years 1990-93 brought about fundamental changes on both fields of social interaction in state-owned enterprises: industrial democracy (participatory type of management) and the trade unions.

The workers' councils, previously the principal form of industrial democracy, were in a deep crisis. In times when reforms of real socialism were attempted, especially in the second half of the 1950s and in the 1980s, workers' councils, later on also called employees' councils, gained a reputation as the leading force opposing totalitarianism in the economy, the exponent of the social aspirations of the employees, and a stimulator of micro-scale efficiency adjustments. After 1989, however, under the radically changed circumstances, they have become 'structures from a different era', clearly maladjusted to the new conditions and needs. Critics argue, not without reason, that the councils in their existing shape threaten to weaken the motivation for efficiency, interfere with rational decision-making, and stand in the way of efficient management. There are two reasons for that. The first one is the position of the councils in the system of governance in the enterprise, which, legally, grants them an equal, or, in some cases, even superior status to that of the director. Secondly, the workers' self-management used to be incorporated into a highly detrimental economic system under which state-owned enterprises had to function and which the post-communist governments failed to modify to a satisfactory degree over a period of four years. After 1989, and especially in the years 1991-92, factors such as the economic crisis in many state-owned enterprises without a clear prospect of recovery; forced privatization, seen as the only reasonable path of transformation; a new situation in the trade unions, and also the climate of antipathy or even downright hostility towards self-management created by the authorities and many political parties, have all considerably weakened the workers' councils. In a growing number of enterprises, self-management has in fact ceased to play any significant role.

The actual influence of self-management on the adjustment of enterprises to the market economy and on their organizational and technical restructuring awaits detailed investigation and a balanced evaluation. Some studies seem to indicate that at the beginning of the transformation in 1990, self-management was in many cases an active participant, playing a beneficial role.[8] Later on, due to the factors mentioned above, this kind of involvement became reduced, and self-management took in several cases an increasingly conservative and pro-consumption stand.[9]

Similar observations apply to the attitude of self-management towards ownership transformations in enterprises and their privatization. There certainly have occurred instances of efficiency-stimulating transformations being blocked by workers' councils. But in many cases self-management did not put any obstacles in the way of the necessary reforms initiated by the authorities or the management, including the transformation of enterprises into Treasury-owned commercial-code companies, which opened up the way to full privatization. Frequently, it was self-management who actually postulated and advocated this kind of reform.[10] It was only natural for self-management and its activists to support the idea of employee-ownership and, whenever this was feasible, employee partnerships and leasing.[11]

Very little is known about the actual role of the greatly modified form of participatory management which functions in those traditional state-owned enterprises which have been transformed into joint-stock companies. Namely, employees have the right to appoint one third of the board of

126

directors, as long as the Treasury controls more than 50 per cent of the stock. One can presume, however, that since the position of the board of directors in companies controlled by the Treasury is in general weak, employee representation plays in them only a limited role, or no role at all.

In 1990-93, the characteristics of trade unions in state-owned enterprises included:

1   full autonomy;

2   pluralism of organizational forms, which mostly found an expression in the rivalry between organizations belonging to the All Polish Trade Union Alliance (OPZZ) and to Solidarity; and

3   decreasing employee participation.

Under such circumstances and in view of the deepening recession, the unions concentrated on advancing claims for the protection of jobs, preservation of the real-wage level, and other benefits. This often led to disruption of public relations and hampered the otherwise possible and economically justified structural changes in enterprises, in the fields of ownership, technology and organization. It seems, however, that in 1992, and especially since its latter half, the attitudes of the unions underwent a certain evolution. The continued pressure for efficiency of enterprises, not yet sufficiently strong, perhaps, but with a tendency to increase, produced in the unions a sense of shared responsibility for the fate of their enterprise, which became increasingly dependent on its performance, creativity, and profitability. Therefore, the priorities of union leaders became inevitably influenced by the criteria of micro-scale efficiency and long-term prospects of the enterprise in the competitive, market environment.[12]

Financial management

The reorganization of financial management along rational lines commenced in the 1980s was continued. It focused on issues such as taxation, the overall relationship between enterprises and the state budget, the distribution of profit, and capital funds. A large proportion of the changes was connected with the differentiation of state-owned enterprises according to their status, in connection with the formation of companies wholly or partially owned by the Treasury. The changes were limited in scope and not very radical, either, and they did not make it possible to rectify the inherited defects of the system and meet the challenge of the transformation. Three fundamental issues remained unsettled:

1   equalizing in a reasonable way the tax burden placed on private and state-owned enterprises;

2   tough budget constraints, connected with the problem of corporate self-financing;

3   finding efficient, but economically sound, ways of dealing with the widespread heavy indebtedness and insolvency of enterprises.

The inequality of taxation consisted first of all in the existence of two kinds of payments in favor of the State Treasury from which private firms were exempt. These were and still partially remain:

1   the fixed assets tax (so-called *dywidenda*) in traditional enterprises, or interest on capital paid by Treasury-owned commercial-code companies; and

2   the progressive (penalty) tax on wage increases (so-called *popiwek*) in traditional enterprises and companies in which the state holds more than 50 per cent of the stock.

However, it must be emphasized in this context that widespread assumptions about financial discrimination of state-owned enterprises is far from obvious and needs some qualification.[13]

Let us observe, then, that the fact that some of the profit earned by state-owned enterprises is taken away by the Treasury as their owner is nothing else than normal execution of property rights, which, obviously, occurs in the private sector, too. On the other hand, both the *dywidenda* payment in its present form, and the interest on capital paid by Treasury-owned commercial-code companies, represent fixed charges of the nature of taxes, which depend neither on the profitability of the company, nor on its investment needs. Compared with the owner's dividend in a typical commercial code company, this is indeed a kind of burden highly detrimental to enterprises. In this sense and to this extent only can one really speak about some form of discrimination.[14]

The case of the *popiwek* tax is a far clearer instance of discrimination, but it is not all that simple, either. It is a fact that since 1990, the *popiwek* has had no equivalent in the private sector. In this sense, it is unquestionably an excessive burden. It is likewise obvious that the *popiwek* has in some ways a negative influence on the efficiency of enterprises, even though its principles and scale are being constantly adjusted. For a large proportion of both the employees and the managers in the public sector, the *popiwek* is a symbol of the injustice and harm to which the sector has been exposed. On the other hand, however, there is no refutation of the argument raised by the authorities, namely, that in view of the still strong and unilateral orientation of state-owned enterprises towards the maximization of wages, in other words their limited ability to conduct a rational wage policy, no instrument can replace the *popiwek* in the economic policy. On such interpretation, the tax appears to be one of the 'anchors' of economic stability. One can thus presume that what the authorities could possibly be blamed for in connection with the controversial *popiwek* is not the very fact of its use under the existing circumstances, but rather the lack of sufficient efforts on their part to alter these circumstances.[15]

Another aspect of the inequality of sectors was the greater facility with which the budget could collect its revenues due from state-owned enterprises. According to the Central Planning Office data, the total revenue yielded in the first nine months of 1992 by the public sector was three times higher than the corresponding figure for the private sector.[16] One could hardly accept the explanation of this difference by the allegedly far greater profitability of state-owned enterprises.

Soft budget constraints, which used to be a typical anti-efficiency feature of state-owned enterprises under real socialism, became significantly reduced in volume in the years 1990-93, but not brought down to acceptable dimensions. All main kinds of state subsidies for enterprises were rapidly reduced (with due allowance for inflation) during that period, both in relative and absolute terms, which meant considerable progress in this field. Unfortunately, this was not the case in other areas: there occurred problems with the settlement of liabilities to the budget, suppliers and banks, but also cases of disrespect for the rights of enterprises to exercise full control of the assets placed at their disposal. The overdue payments owed by the enterprises to the highly unbalanced budget ran into thousands of billions of zlotys: towards the end of 1993 they amounted to nearly 100,000 billion, of which sum presumably more than a half was accounted for by state-owned enterprises. There had accumulated an enormous volume of unpaid debt owed to suppliers - in other words, forced mutual credit among economic entities, also known as cash-flow disturbances - which exceeded at the end of 1993, according to the data of the Central Statistical Office, the amount of 279,000 billion zlotys. Moreover, the fact that many enterprises fail to meet their financial obligations to the budget and suppliers does not mean that they use the money to pay off their debts to the banks. Conversely, the problem of overdue bank loans is becoming more and more acute. According to the estimates of 14 leading Polish banks, about 2,100 state-owned enterprises had lost their credit capacity by the end of February 1993. The reason was their poor financial performance. Out of the 6,963 public units investigated by the Central Statistical Office, 22.5 per cent measured by the share in total income showed losses in their gross balance for 1993. This means they were unable to cover their expenses (including the value added tax and customs duties) from their income. As against 1992 and 1991 the situation has improved partially in this respect, notably due to overall economic recovery.

In view of such a crisis, the number of state-owned enterprises that undergo liquidation owing to their poor performance or go bankrupt appears to be relatively small. In the period 1991-93, liquidation proceedings were instituted due to poor performance in 1,119 enterprises, and bankruptcy proceedings, in 378.[17] Let us add that these were mostly small and medium-size enterprises. Large enterprises which sustained losses and were unable to meet their obligations usually could survive, owing to the 'lenience of creditors', and sometimes also by converting into cash the public property placed at their control, without formally breaking the law! This is how the pressure for efficiency - by all means a desirable thing - becomes diluted, particularly in the case of large enterprises, by the relaxation of their budget constraints. Practically all the participants in the microeconomic game have contributed to such a state of affairs: agencies of the Treasury - mostly for socio-political reasons; suppliers and banks - due to some specific factors of economic conditioning, and sometimes also by yielding to political pressure; and the enterprises themselves - by taking advantage of the possibility, offered by the system, to finance losses from public capital and extensively use political pressures.

The need to open up special paths for state-owned enterprises to offer them a way out of their financial crisis results from the specificity of the situation: massive scale is not the only important property of the phenomenon. Apart from the general factors, such as recession, liberalization of import policy, or high interest rates, the problems are also due to the special handicaps imposed on enterprises, so to say, from the outside, by the past and present political decision-makers. In terms of real conditions, these include the technological and organizational deficiencies, as well as often high indebtedness, inherited from the communist economy - and in terms of the system or regulation, the antiquated operating principles, which have persisted without any radical reform, affecting the performance of state-owned enterprises. It is obvious that the strategies of overcoming the crisis - if they are to promote transformation instead of blocking or impairing it - have to be effective, but, on the other hand, they should also meet the tough criteria of economic rationality. Satisfying both these requirements, which in a sense run against each other, appears possible; however, under the assumption that grass roots energy, inventiveness and initiative are to a high degree activated and directed towards efficiency improvement.

The first three years of the transformation failed to produce such strategies. There were many reasons for that, presumably, in no small part of a doctrinal nature. The authorities responded to the financial crisis of numerous state-owned enterprises by instituting liquidation and bankruptcy proceedings in smaller economic units, and relaxing the budget restrictions applied to larger ones, especially in cases when their liquidation was viewed as too risky, due to the unemployment threat or political tensions. In the second half of 1992, the government of Hanna Suchocka undertook legislative steps and a public debate in order to work out a recovery procedure better suited to the need of the situation, called bank arbitration. In February 1993, the Parliament accepted this arrangement, passing the law on financial restructuring of enterprises and banks, popularly called the 'debt-settlement law'.[18]

Bank arbitration means simplified and fast proceedings, a broad range of recovery procedures, involving both the debtor and the creditor, like capitalization of interest, conversion of liabilities into stock, or new loans or guarantees extended to the debtor. It presupposes active participation of the Treasury agencies. In each concrete case, the settlement should be based, of course, on a convincing recovery program for the enterprise concerned. The institution of amicable settlement is supplemented by two important complementary mechanisms: the possibility of public sales of debts owed to banks and the obligation of the State Treasury (the Ministry of Finance) to increase the capital of the state-owned banks participating in the settlement scheme, by way of cash transfers and issuing of special Treasury bonds.

The new recovery procedure seems to be a rather sound and convincing solution. Still one would have to wait and see how it works in practice.[19] Presumably, the success of this undertaking will mainly depend on:

1 genuine involvement of the State Treasury; and

2 a consistent policy in other relevant fields, correlated with the debt settlement program.

The first condition has to do with issues like a sensible policy on the remission of payment arrears due to the budget, availability of economically justified credit guarantees, or even, in some cases, gradually diminishing subsidies, and the amount of extra capital required by banks for this operation, for which external sources of funding would have to be provided.[20] The second point, in turn, concerns the necessity of linking the recovery procedure with tough restrictions on financing enterprises by the budget (the requirement of unshakable self-financing), and some other changes in their mode of operation.

External environment

Earlier in this section, the ongoing changes in the environment of state-owned enterprises have been called a breakthrough. Their essence was the rapid transition to market economy. Liberalization of prices, balancing of supply and demand, and abolition of the command system in the economy turned enterprises (apart from those of 'special importance', i.e. in the energy industry, defense industry, some branches of transportation, communications, and municipal services) into economic actors enjoying an increasing allowance of decision-making autonomy, controlled by the

131

maturing market forces. The so-called internal convertibility of the zloty and radical liberalization of foreign trade and other forms of foreign economic contacts brought enterprises into the world markets and exposed them to the mechanisms and tendencies that prevail there. These processes had many special features, some of which found very concrete expression in the ways the developing market environment influenced the enterprises. The most important of these include the following:

Changes in the composition of agents active on the market. These comprise first of all rapid development of private firms, either domestic, or involving foreign capital, and the relatively rapid privatization of the economy effected in this way. During the four years from 1990 to 1993, the number of private companies increased by 55,500, the number of joint ventures (with foreign capital) by 14,600,000, and privately-owned small businesses by 911,900,000. The share of the 'new' private sector (i.e., including cooperative businesses) in the gross domestic product grew from 28.6 per cent to over 50 per cent (these figures do not take into account the 'grey' sector), and in the overall employment statistics, from 47.2 to nearly 60 per cent. Of course, these indicators continue to increase. In the majority of branches of the economy, the private sector has gained a significant share, and in some (trade, construction, imports, many types of services) it already boasts a dominant position.

Recession and signs of recovery. The precipitous drop of economic activity that occurred in Poland during the first two years of the transformation meant for the enterprises first of all decreased demand and mounting problems with selling goods and services. The gross domestic product dropped in Poland by 17-18 per cent during 1990-92 (once again, ignoring the 'grey' sector), industrial output by about 30 per cent, and agricultural production by about 15 per cent. In the years 1991-93, recession struck most economies in Western and Eastern Europe, including the German markets, which are of special importance for Polish enterprises. There followed massive unemployment and dramatic changes on the job markets. In Poland, the total number of unemployed registered in March 1994 amounted to 2.9 million, or 16.0 per cent of the professionally active part of the civilian population. However, since the beginning of 1992, the Polish economy has been showing signs of recovery. Overall employment, as well as employment outside agriculture, has gradually ceased to drop, thus growth of unemployment has been checked, and industrial production, the volume of construction works and many types of services have started to grow (despite some fluctuations). Gross domestic product increased in 1992 by about 1.5 per cent, in 1993 by about 4 per cent, and in 1994, a growth of 4.5 per cent is expected.

High inflation. Although hyper-inflation was successfully coped with at the beginning of 1990, efforts to bring it further down to a level that does not interfere with the efficiency of market mechanisms ran up against serious obstacles. Taking the prices of consumer goods and services as the indicator, inflation was 70.3 per cent in 1991, 43 in 1992, 35.3 in 1993, and in 1994 is expected, according to the official estimates, to be about 27 per cent. Its harmful effects on enterprises include, first of all, the following:

1 a high uncertainty level, especially when medium and long-term decisions are concerned;

2 extremely high interest rates on investment credits (linked with the inflation rate and amounting to an average of 40-50 per cent p.a. in the years 1992-93), which precludes any wider access to such credits;

3 the 'credit pitfall', in which many enterprises which had run into debt in the 1980s or failed to adapt to the dramatic changes in the fiscal and credit system, got entrapped.

Limitations on export and import other than rates of exchange. They usually take the form of quotas (ceilings) set up by an administrative decision, or discriminating and prohibitive duties. Such regulations can be justified by the necessity to protect domestic manufacturers, but they slow down the market transformation of the economic environment of enterprises, hamper rationalization of the price structure, and weaken the pressure for efficiency, hence impeding the desired structural transformations. Particularly harmful in this respect are the above-mentioned barriers blocking access to the markets of the EU, which has become since 1990 the main economic partner of Poland and the main area of expansion for Polish exports. In view of the disintegration of the Comecon and the critical economic situation of the former Soviet republics, their markets, which used to play a key role for many Polish exporters, have lost much of their importance or simply have been closed to Polish exports.

Underdevelopment of the financial infrastructure. Although many private banks have been founded (over 100; usually of a modest size), and privatization of state-owned banks, dominating the market, has begun, commercial banks remain rather inactive. They tend to avoid taking risks and prefer safe investments, like Treasury bonds and other state securities, which yield a relatively high interest. This is connected with the high budget deficit, persistent since 1991, which is being financed, increasingly, by emission and sales of securities on the tight Polish market. The main institution of the emerging capital market in Poland, the Warsaw Stock Exchange, has shown an exceptionally rapid increase in the turnover and prices during 1993 and the first two months of 1994. At the beginning of March 1994 the market steeply collapsed, and its plausible recovery, presumably, could be only partial and time-consuming. The market is still very narrow, dominated by short-term speculation. In May 1994, only 24 companies were formally quoted, and most of the transactions were secondary dealings in stock issued by the Treasury in the course of privatization.

Instability and unpredictability of the regulations. These concern various areas, but especially the tax system and foreign trade. Changing regulations are to some extent a natural element of the systemic transformation. This, however, is combined with a number of current political problems: mutability and weakness of government, different forms of political pressure, and social tensions. For enterprises, the instability, and especially unpredictability of the regulations remain a major cause of incertitude which hampers the rationality of decision-making.

To sum up, one can say that the external environment has been influencing enterprises as a powerful stimulant of efficiency, but also as a major inhibiting factor. The first type of influence has to do with greater autonomy, a more rational price structure, and, most important of all, competition, which induces efficiency. The changing environment made efficiency-oriented adjustments at the same time possible and necessary, not only in order to expand, but also simply to survive. The other, negative group of factors is connected with the high uncertainties of the situation, and difficult access to both capital and larger markets. Conclusion: the performance of state-owned enterprises has been dependent in complex ways both on the features of their environment and on their internal principles of operation, discussed above.

Financial standing of enterprises: crisis and signs of recovery

The relative prosperity of 1990 was followed by a deep financial crisis of enterprises in 1991-92. Towards the end of that period and in 1993, growing signs of recovery became visible. During all these years, the financial standing of enterprises was very strongly differentiated both within and across branches, which was due to a whole range of factors, both historical and contemporary.

Despite the general drop in overall economic activity, 1990 turned out to be relatively good to the traditional-type publicly-owned enterprises. The main financial indicators for the sector and its branches were lower than the year before, but still on the positive side, and relatively high, at that. The general rentability rate (the ratio of the gross financial result to the cost of total income acquisition) amounted to 29.9 per cent, and the profitability index (net financial result compared to the total income) to 11.3 per cent. Enterprises achieved this success thanks to the unique combination of very special factors, such as: abrupt devaluation of the zloty, re-evaluation of material and foreign-currency reserves, and a high volume of profitable exports to the Soviet Union, financed largely by state funds. The moment of truth came in 1991, when the rentability rate dropped to 4.8 per cent, and the profitability rate to minus 1.3 per cent. The escalation of crisis of enterprises in the years 1991-92 and the ensuing signs of recovery towards the end of that period and in 1993 are illustrated in the table below.

Table 5.1
Some economic indicators pertaining to enterprises in 1991-93
(%)

	1991		1992		1993	
	$1\text{-}6^a$	$1\text{-}12^b$	1-6	1-12	1-6	1-12
Rentability rate	6.7	4.8	3.3	2.2	3.6	2.9
Profitability rate	0.1	-1.3	-0.9	-1.5	-0.1	-0.5
Income of enterprises that sustained losses in their gross balance compared to total income of enterprises	19.0	20.8	24.2	23.7	23.1	24.0
Credits and liabilities compared to money, dues and claims (end of period)	114.4	134.7	140.7	159.0	165.5	165.9

a   January - June
b   January - December
Source: Biuletyn Statystyczny GUS (Statistical Bulletin of the Central Statistical Office)(1992), vol. 8, pp. 65, 72-73; (1993), vol. 5, pp. 71, 79, 83; and (1994), vol. 3, pp. 75, 80, 90.

Towards the end of 1992, the escalation of the crisis began to slow down, and the first months of 1993 brought some symptoms of recovery: both the rentability rate and profitability rate were better than in 1992. However, there is still a long way to go to overcome the financial crisis in enterprises, especially state-owned ones. According to the estimates of 14 leading commercial banks referred to above, 4,666 enterprises, including about 2,100 in the public sector, had lost their credit capacity by the end of February 1993, which puts them under a threat of bankruptcy or liquidation.

Let us observe, besides, that the figures given above present a highly inaccurate picture of changes in the financial standing of state-owned enterprises, one that only indicates the main tendencies. The data represent statistical aggregates calculated for all the enterprises (with the exception of agriculture and forestry), state-owned and private alike, with an employment of 50 or more in industry and construction, and 20 or more in the remaining branches. State-owned enterprises are the dominant component of those aggregates, but their contribution to the overall picture varies from branch to branch and has a steady tendency to drop. There are many facts and a lot of observational data suggesting that the financial standing of state-owned enterprises was generally worse than that of private ones. Official statistical materials, however, do not corroborate this opinion. A comparative study done by the Central Statistical Office (CSO, 1994), involving 6,963 publicly-owned economic entities and 12,481 private ones, indicates that the rentability rate of state-owned enterprises in 1993 was much higher, at 3.9 per cent, than the respective aggregate figure for private enterprises, amounting to a meagre one per cent. In terms of profitability rate, private firms did better than state-owned ones on the whole (-0.5 and -0.4), and in all distinguished branches except domestic trade. A similar study (CSO, 1992) covering the first eleven months of 1992 yielded comparable results, albeit with a much higher supremacy of the public sector.[21] Still, I do not think these data warrant any far-reaching generalizations. There are probably many different factors accounting for the disparity in the rentability index. Presumably, they have little to do with the efficiency of enterprises as it really is. Nevertheless, these are surprising results, which point to the need for more profound inter-sector studies.

In general, the situation of state-owned enterprises in the years 1991-93 was very difficult and had many features of a crisis, but at the same time it varied greatly depending on the sector, branch, and many other factors. This is demonstrated by the CSO materials quoted above, and also by some in-depth analytical studies, e.g. the one carried out by the Institute of Economic Sciences of the Polish Academy of Sciences, comprising an analysis of the 500 largest enterprises in the manufacturing industry in 1991 and 1992,[22] the investigation of Chmiel from the Statistics and Economics Research Unit run jointly by the Central Statistical Office and the Polish Academy of Sciences,[23] and the study by Gomulka on the standing of enterprises in different ownership sectors between January and September 1992, including over 7,000 state-owned enterprises.[24] Using a certain synthetic criterion of financial standing assessment, Gomulka rated the standing of 29.0 per cent of state-owned enterprises as very bad, 50.8 per cent as satisfactory, and 20.2 per cent as good or very good. Thus the overall picture points at the same time to a serious crisis, affecting nearly one third of the population under study, and to successful adjustment, owing to which the majority of state-owned enterprises (over two thirds of the population) managed to build up the financial resources needed for survival, or even expansion, under the condition of recession and an increasingly difficult situation on the market.

Irrationality of SOE behavior

The crisis affecting a large part of publicly-owned enterprises is a consequence of a number of interdependent factors, three of which appear to be the most important. These include, beyond the deep recession:

1 outdated and depreciated production capital, of awkward structural composition, combined with obsolete technologies;

2 liberalization of the economy and its exposure to external influence and competition, which it is in many cases unable to cope with;

3 irrational behavior of enterprises, from the point of view of efficiency, which provokes now much criticism, usually justified. The last report on the state of the economy prepared by the outgoing Bielecki - Balcerowicz government says that:

> ... most state-owned enterprises assumed in the years 1990-91 a passive attitude, devoting little attention to development, and concentrating on the current problems, instead. At the same time they were engaged in an aggressive dispute with the government, demanding higher wages, tax relief, preferential credits, and an increase/decrease of customs duties - depending on how a given enterprise perceived its interests.[25]

Such a critical evaluation was confirmed by numerous studies, carried out by researchers from Poland and abroad, who usually emphasize the wage-orientedness of state-owned enterprises and their tactic of waiting out the new attempt to reform the economy along market lines.[26]

In the first years of the transformation, the irrationality of behavior of many, or even most, enterprises was in a way an objective consideration. After all, it had been engendered by the experience of real socialism, its failed reforms, and the resulting habits and reaction patterns of employees and managers. Moreover, such types of behavior were often justified or reinforced by the day-to-day practical experience of the transformation the way it was. What is meant here is the difficult-to-overcome, anti-efficiency remnants in the organizational and economic make-up of enterprises, which correlate with the ways enterprises are managed by the government's agencies.

The market-oriented changes in the functioning of enterprises, especially in their economic environment, have been stimulating a gradual, positive evolution of their behavior. This is convincingly demonstrated especially by the most recent studies and analyses, both economic and sociological.[27] It is likely that this trend will gain momentum in time. The overcoming of recession and the first signs of economic recovery were certainly due to the dynamism of the private sector. One must not forget, however, the importance of the role played in this process of the adjustments made by many state-owned enterprises.

But in view of the persistent, profound inadequacy of the internal operating principles of state-owned enterprises, and institutions of owners' supervision, pressure from the outside was not enough to bring about a complete change. One could say that many of the anti-efficiency types of behavior persist not because they could not be eliminated, but because of a policy which operates with limited instruments and means. This observation is of fundamental importance for the evaluation of the performance of enterprises until now and their chances for the future. The conviction that the reform of the internal operating principles in state-owned enterprises is inconsequential, together with the belief that the desired improvement of their efficiency can be achieved through outside pressure alone, had become - as has been said before - part of the economic philosophy of the first two post-communist governments. Hitherto experience has compelled one-time adherents of such a policy, and even some of its creators, to admit its delusiveness and harmful consequences.

Some hints of this can be found already in the late 1991 report on the state of the economy, quoted above, and later in various materials prepared by the next governments. One could hardly express this idea with greater clarity than was done by Hanna Suchocka's government in the Outline of the Pact on the State-Owned Enterprise under Transformation of August 1992. It says:

> State-owned enterprises exist, basically, in the same form they were established in the centrally controlled economy ... Enterprises which were formed under the conditions prevailing in those days are now unable to cope with the conditions of market economy ... In the transition to market economy, they were left to themselves.[28]

Conclusion

The analysis presented above allows us to conclude that the weak points that persist in the functioning of state-owned enterprises and their maladjustment to the requirements of efficiency are largely due to their internal principles of operation, which badly need a radical reform. The main problems here are connected with:

1   the imperfect and conflict-prone organizational structure and distribution of the decision-making powers, on both micro and macro economic levels; and

2   certain ineradicable remnants of the past in the sphere of the credit system, relationships with the budget, and other areas of finance, which restrict the autonomy of enterprises and counteract the pressure for efficiency.

On the other hand, the external environment of enterprises, the beneficial influence of which I have repeatedly stressed, provides only a somewhat limited market-type stimulus for efficiency. This results from the high inflation rate, underdeveloped financial structure, barriers in foreign trade, and lack of sufficient stability of the regulations instituted by the state.

The fact that the problems responsible for the undesirable types of performance of state-owned enterprises have not yet been addressed, can and should be treated as an important reserve of efficiency that has yet to be utilized. Making this reserve productive does not require, in most cases, any major capital investment. What it does require is rejection of dogmatism, and implementation of a comprehensive, efficiency-oriented systemic reform of publicly-owned enterprises. Hopefully, the Pact on the State-Owned Enterprises under Transformation, the product of tripartite negotiations, could provide an important step in that direction, once some of the beneficial arrangements it postulates are put into practice and developed.

Notes

1. Council of Ministers (October 1989), *Program Gospodarczy. Główne zalozenia i kierunki*, Warsaw, pp. 4-5.
2. Minister of Finance and President of the National Bank of Poland (22 December 1989), *Letter of intent* (Polish version), Warsaw, p. 21.
3. According to data published recently by the Central Statistical Office (CSO) the profitability rate (relation of gross financial result to cost of total income acquisition) of all Treasury-owned stock companies in the period January-November 1993 amounted to 2.2 per cent. In the same period that rate of all publicly-owned enterprises was higher, namely 5 per cent, and of all already privatized companies through the so-called capital procedure 7.6 per cent.

Informacja o sytuacji spoleczno-gospodarczej kraju, Rok 1993, GUS 1994-01-27, p. 70.

4. The Law of 13 July 1990 on privatization of state enterprises, *Dziennik Ustaw* (1990), no. 51, and (1991), no. 60.

5. The Law of 30 April 1993 on national investment funds and their privatization, *Dziennik Ustaw* (1993), no. 44.

6. In 142 employee partnerships formed during 1990 and 1991 the average share of managers already at the outset amounted to 27.2 per cent and afterwards was increasing gradually. Olgo-Bagienska, Teresa et al. (1994), 'Spólki pracownicze lepsze od innych, ale ...', *Zycie Gospodarcze*, no. 12.

7. There are prospects of probably not distant general institutional overhaul connected initially with already mentioned common (mass) privatization. Its program stipulates the formation at the end of 1994 of a dozen modern holding organizations named National Investment Funds (NJFs). At the first stage of the program NJFs will take over the majority of Treasury-owned shares in about 550-600 public stock companies. It is assumed that during approximately two years the NJFs will operate as public intermediary institutions, and afterwards will be privatized through semi-free distribution of their shares among all adult Polish citizens. In addition to that the present leftist government has promised to radically restructure the network of all remaining institutions performing Treasury's proprietary functions. Central Planning Office (23 March 1994), *Program polityki gospodarczej na lata 1994-97*, Warsaw, pp. 58-60.

8. See among others: Dabrowski, J. M., Levithas, A. and Fedorowicz, M. (1991), *Przedsiebiorstwa panstwowe w roku 1990. Wyniki badan*, Institute of Studies on Market Economy and Property Rights, Warsaw.

9. See: World Bank Resident Mission (1992), *Microeconomics of transformation in Poland. A survey of state enterprise responses*, Warsaw. And also: Panków, W. (1992), 'Autodestrukcja reformy', *Przeglad Spoleczny*, no. 5.

10. This was shown among others in the empirical study of Gilejko, L. (1991), *Raport z badan nt.: uwarunkowania strategii organizacji i pracowniczych ruchów spolecznych wobec prywatyzacji*, Warsaw; and then fully confirmed by the Ministry of Ownership Transformations (1993) in its report on privatization from 1990 to 1992, *Raport o prywatyzacji 1990-92*, Warsaw, p. 18 and p. 93.

11. Ibid., pp. 45-49, and Jarosz, M. (ed.) (1994), 'Spólki pracownicze', *ISP PAN*, Warsaw, pp. 79-84.

12. This tendency was ascertained in several recent empirical studies, among them carried out by Professor M. Jarosz and her team that investigated the social attitudes and responses in the 500 biggest state-owned manufacturing enterprises. Jarosz, M. (ed.) (1993), 'Prywatyzacja. Szanse i zagrozenia', *ISP PAN*, Warsaw, pp. 113-114.

13. In 1993, 1.5 per cent of total budgetary revenue came from dividends and interest on capital paid by public enterprises, and 2.2 per cent from tax on increase of their wage and salary (*popiwek*).

14. The above mentioned Pact on State Enterprise in Transition stipulates that the dividends and interest on capital should be unified and become more flexible,

which means that each enterprise would be able to choose one of the alternative and more relational forms of this payment.

15. Because of the President's veto the new liberalized law on *popiwek* so far did not come into force. Thus since April 1994 state-owned enterprises do not pay this tax. The Pawlak government is determined to try again, hoping that the law, slightly changed, under the pressure of economic logic, will be approved finally by both the Parliament and the President in the near future.

16. Central Planning Office (1993), *Wstepna ocena sytuacji spoleczno-gospodarczej w 1992 r.*, Warsaw. Similar evaluations have been presented by the CSO.

17. Central Planning Office (1994), *Sytuacja Spoleczno-Gospodarcza 1993 Roku*, Warsaw, p. 57.

18. The Law of 3 February 1993 on financial restructuring of enterprises and banks, and on changes of same laws, *Dziennik Ustaw* (1993), no. 18.

19. Till the end of March 1994 about 180 mostly large public enterprises were covered by this procedure, and 70 debt settlement contracts have been concluded. See: Badzio, P. (1994), Gorycz Wierzycieli, *Zycie Gospodarcze*, no. 12. The sharply differentiated opinions on this matter were presented in an editorial: 'Test na koszenie', *Zlycie Gospodarcze*, 1994, no. 15.

20. This condition seems to be met to a high degree through a set of complementary regulations and economic policy decisions.

21. For the total population rentability rate amounted to 4.3 per cent in the public sector, and to 0.1 per cent in the private one. Corresponding figures concerning profitability rate were -0.4 and -1.0.

22. Balcerowicz, E. and Macieja, J. (eds) (forthcoming 1994), *Polska elita przemyslu przetwórczego w 1991 i 1992 r.*, Institute of Economic Sciences, Polish Academy of Sciencies, vol. 1 and 2, Warsaw.

23. Chmiel, J. (1993), 'Financial situation in the enterprises' in Zienkowski, L. (ed.), *Polish economy in 1990-1992. Experience and conclusions*, RECESS, Warsaw.

24. Gomulka, S. (1993), *Sytuacja finansowa przedsiebiorstw polskich w r. 1992 i jej wpływ na polityke monetarna i fiskalna*, MS, Warsaw. The next edition in English (1993), *The financial situation of Polish enterprises 1992-93 and its impact on monetary and fiscal policies*, CASE, Warsaw.

25. Council of Ministers and its Economic Committee (1991), *Raport o stanie gospodarki*, MS, Warsaw, p. 4.

26. Among others: The Institute of Studies on Market Economy (1993), *Raport z Badan*, MS, Gdansk; Caban, W. (1991), 'Raport z badan: funkcjonowanie przedsiebiorstw panstwowych w procesie urynkowienia gospodarki. Zeszyty sygnalne', INE PAN, no. 56, Warsaw. World Bank Resident Mission (1992), Microeconomics of transformation in Poland: a survey of state enterprises responses, MS, Warsaw.

27. Pinto, B., Belka, M. and Krajewski, S. (1993) 'Transforming state enterprises in Poland: evidence on adjustment by manufacturing firms', *Brooking Papers on Economic Activity*, vol. 1. Baczko, T. (ed.) (1993) 'Microeconomic foundation of transition', *INE PAN*, Warsaw. Jarosz, M. (ed.) (1993), Prywatyzacja. szanse i zagrozenia, *ISP PAN*, Warsaw.

28. Council of Ministers (1992), Zalozenia paktu o przedsiebiorstwie panstwowym w trakcie przeksztalcania, MS, Warsaw. See also: (1992), Pakt o przedsiebiorstwie panstwowym, *Prywatyzacja*, no. 9.

References

Aslund, A. (1993), *Wnioski z pierwszych czterech lat zmian systemowych w Europie Wschodniej*, CASE, Warsaw.

Balcerowicz, L. (1993) 'Demokracja nie zastlpi kapitalizmu', *Przeglld Polityczny*, special volume.

Balcerowicz, L. (1992), *800 dni. Szok kontrolowany*, BGW, Warsaw 1992.

Bugaj, R. and Kowalik, T. (1992), 'The privatization debate in Poland', in *Privatization in Eastern Europe*, Reiner Institut, Vienna.

Central Board of Planning, *Quarterly and yearly reports on the socio-economic situation in Poland; 1990-94*, Warsaw.

Central Statistical Office (1993), *O sytuacji spoleczno-gospodarczej kraju, rok 1992*, p. 22.

Central Statistical Office (1994), 'Wyniki finansowe przedsiebiorstw w 1993r' in *Informacja o sytuacji gospodarczej kraju, luty 1994r*, p. 29, Warsaw.

Central Statistical Office, *Yearbooks; 1990-93*, Warsaw.

Central Statistical Office, *Monthly Statistical Bulletins*, 1991-94, Warsaw.

Central Statistical Office, *Quartely and yearly reports on the socio-economic situation in the country; 1991-94*, Warsaw.

Council of Ministers (1993), *Raport o stanie panstwa*, Warsaw.

Dlbrowski, M. (1991), 'Od gospodarki planowej do gospodarki rynkowej. Tempo i etapy transformacji systemowej', *Studia i Materialy INE PAN*, no. 33.

Dlbrowski, M., (1993) 'Bilans trzech lat', *Przeglld Polityczny*, special volume.

Dreg, Z. and Indrakiewicz, J. (1992), *Swiadomolsc pracodawców a zmiany ustrojowe w Polsce*, F. Ebert Foundation, Warsaw.

Gabrisch, M. et al. (1992), *Advanced reforming countries might reach end of recession*, WJJW, Vienna.

Gomulka, S. (1992), 'Polish economic reform, 1990-91, principles, policies and outcomes', *Cambridge Journal of Economics*, 1992, no. 16.

Gomulka, S. (1994), *Budget deficit and inflation in transition economies*, CASE, Warsaw.

Kotowicz-Jawor, J. (ed.) (1992), 'Procesy dostosowawcze przedsiebiorstw przemyslowych do mechanizmu rynkowego. Raport z badan'. *Gospodarka Narodowa*, no. 5.

Kawalec, S. (1989), 'Privatization of the Polish economy', *Communist Economies*, vol. 3.

Kolodko, G.W. (1992), *Transformacja Polskiej gospodarki. Sukces czy porazka*, BGW, Warsaw.

Lewandowski, J. and Szomburg, J. (1989), 'Property reform as a basic for social and economic reform', *Communist Economies*, vol. 3.

Lipton, D. J. and Sachs, J. (1990), 'Creating a market economy in Eastern Europe. The case of Poland', *Brooking Papers on Economic Activity*, vol. 1.

Mujzel, J. (1993), Przeksztalcenia wlasnolciowe w Polsce (1990-92), Poltext, Warsaw.

Mujzel, J. (forthcoming), *Poland's post-communist transformation and the problems of recession*.

Mujzel, J. (forthcoming), *Privatization in Poland - its achievements, weakness and dilemmas*.

Polish Economic Association (1991), *Drogi wyjlcia z polskiego kryzysu gospodarczego*, PTE, Warsaw.

Sadowski, Z. (1991), 'Niezalezne spojrzenie na prywatyzacje w Polsce', *Gospodarka Narodowa*, no. 3.

Winckler, G. (mod.) (1992), *Central and Eastern Europe: roads to growth*, IMF, Washington.

6 Delayed privatization, financial development and management in Romania during the transition

Irina Dumitriu and Teodor Nicolaescu

Introduction

At the end of 1993, after almost four years of reforms, overall economic performance in Romania seemed to be even worse than during the last period of the communist regime. A closer look at the measures introduced until then points out some important 'missing links': out of about 7,000 companies with state participation, only 253 have been privatized (the smallest), there is no real money market, the banking sector is still highly monopolized and state controlled, the domestic currency is not yet convertible into foreign currency at a market rate, and there is no capital market formed yet. Is it possible that these are the causes of the continuous decline in output in the real economy? What is the interplay between the specific conditions of Romania and the slow progress in these fields?

Delayed privatization and slow reform of the financial sector are definitely counterproductive for the success of the transition to the market system, and the case of Romania gives clear empirical evidence in this respect. Assuming that the market system, based on free enterprise and competition, gives a better utilization of resources and thus, progressing economic efficiency and living standards, this paper tries to establish the causes of the delays in privatization and reform of the financial sector in Romania.

The main issues upon which this chapter focuses are:

1 The 'starting point' or 'the legacy of the past' - the features of the Romanian economic and socio-political system during the communist years have induced patterns of behavior and specific economic and financial flows that affect the success of the transformation.

145

2   During the first transition years a series of measures have been introduced and the sequencing and the results of these measures have also contributed, some times in an adverse manner, to the present situation.

3   The design of reform packages, focusing on institutional change and macroeconomic stabilization, sometimes failed to address important microeconomic issues, like corporate governance and company management.

4   The stage that was reached in privatization and financial sector reform is not satisfactory and is a source of bottle-necks for further reforms, thus the necessity of urgent corrective measures.

The conclusion of the chapter is that rapid privatization and development of the financial sector (especially financial intermediation and financial markets) are seriously affecting the structures of economic and political power in society and no progress can be achieved in the absence of strong political will.

Legacy of the past

Romania started its transition to a market economy and a democratic society much later than other countries in the Eastern bloc and from one of the lowest levels. The Romanian economy at the end of 1989 could be characterized as an incoherent structure of different sectors, an excessively centralized social, economic and management system, with an autarchic reactionary pattern, generating inefficient activities and constantly declining living standards.

> Among the so-called Soviet satellites of Eastern and Central Europe prior to the revolutions of 1989, Romania perhaps closest fit the classical archetype of the socialist command economy. Detailed central planning was carried to absurd extremes by Nicolae Ceausescu, the President and General Secretary of the Romanian Communist Party. The economy functioned almost like a single firm, with most agents simply following instructions and with virtually no reliance on individual motivation to secure desired results.[1]

The private sector in the Romanian economy was practically nonexistent, as some workshops and small farms in the submontane area represented less than three per cent of the national assets[2] and even they operated under strict government control (compulsory deliveries, state-controlled prices on the peasant markets).

However, Romania is often considered to have had also two comparative advantages at the starting point of the transition period. First, the fact that it "... was less dependent on Comecon markets and - despite the autarky - relatively more experienced in serving the competitive markets of the West than other East European countries ..."[3] although the Romanian exports to Western countries were not at all competitive, they were 'forced' at all means because of the rush for hard currency. Secondly, at the end of the communist regime, Romania had the lowest debt load, but the efforts made to pay back the foreign debt impaired not only the population's standard of living, but also economic potential, through cuts of imports and reduced investment.

No pre-reform period was allowed to prepare the ground for the transition to a market economy, so that the December 1989 revolution represented an abrupt break with the old system; most of the old rules and institutions were rejected or simply disregarded, thus calling for the repeal of many anachronistic laws and the hasty enactment of new ones. Central planning collapsed while the regulatory framework for a new system was far from complete.

As an example, it is interesting to have a look at the management structure of enterprises. Until 22 December 1989, under Law no. 5/1978 (the Law on 'self-financing and workers' self-management of enterprises'), in economic units in Romania the ultimate decision-maker in an enterprise was the chairman of the 'working peoples' council', that is the secretary (the head) of the communist party organization in the enterprise. In December 1989 the Romanian Communist Party disintegrated and local organizations disappeared. However, Law no. 5/1978 was not explicitly abolished until July 1990, when a new structure was introduced (see sections three and four). So, during the first six months of 1990 we witnessed the transformation of the 'working peoples' councils' into Administrative Councils (somehow similar and acting like boards of directors for the relevant enterprises), that is, self-management at its most.

These special circumstances of Romania could justify in part the delay in the transition process to a market economy system and the adverse results of a quite ambitious reform program started three years ago, somehow similar with the comments based on the case of Hungary in Laszlo Csaba's discussion of crisis-induced reform. Although Csaba argues that "... deliberations about economic reform are always products rather than causes of poor economic performance ...", there may be also more than such a relation.[4]

Transition process: 1990-93

In about three years of transition, Romania witnessed the introduction of a comprehensive reform package, launched in 1990, adjusted several times during 1991 and 1992, and significantly amended in 1993. Reforms were introduced for all parts of Romanian society, but the heart of the transition consists of the economic measures. However, in the fourth year of the transition, results are rather discouraging.

What went wrong? According to the Romanian Cabinet in office at the end of 1993, over the last three years:

> ... the Government's efforts to rely more fully on the price mechanism and on exchange and interest rates to allocate resources in the economy, and to ensure monetary restraint, have been impeded, inter alia, by deep-rooted structural problems including the weakness of financial discipline in enterprises".[5]

These 'deeply-rooted structural problems' are indeed a true cause for the failure of previous macrostabilization programs, but it is nonetheless true that, during the same past three years, the reforms introduced were sometimes inconsistent, not completed for political (and electioneering) purposes,[6] delayed, or their sequencing not seriously considered.

The main components of the Romanian transition measures (from July 1990 to August 1993) have been: institutional change, decontrol and price reform, reform of the administration, commercialization (corporatization) of economic units, privatization, trade policy reform, banking and financial sector reform, industrial restructuring, fiscal policy and tax reform.

Legal environment and institutional change

In the absence of a constitutional settlement (the new Constitution was adopted by referendum only in December 1991), the rights to own real properties and to set up private enterprises were gradually introduced, first by a Decree-law (no. 54/1990, allowing private enterprise with certain limitations) and then by a new Law on Commercial Companies no. 31/1990.[7] A special act was passed to reform state-owned enterprises and to turn them into either commercial entities with the state as shareholder or into 'regies autonomes', a type of state-administered body, with a legal limitation of the amount of subsidization from the central budget and having their budget and financial statements open to public scrutiny (as these are, by law, published as annexes to the state budget). The same act (Law 15/1990) had two provisions that were both a promise and a commitment to privatize (that is to sell a part and freely distribute another) the 'commercialized entities'. Law 15/1990 also stipulated the possibility of long-term leases of state properties, according to special procedures. It is worth mentioning that all these legal provisions did not cover urban real estate properties, nor agricultural estates. The agricultural land has been the object of a separate act, passed only in the spring of 1991. Until August 1991, when the Privatization Act was passed, there has been no mention whatsoever about restitution of confiscated or nationalized properties, and the only legal provision introduced by this act is just another promise that remains to be fulfilled.[8] This fragmented, incomplete and gradually introduced property rights legislation is now among the most serious obstacles to privatization and to foreign investment. We may say that the Romanian legislation on property rights is a 'reformed communist' one, that is permissive (the state recognizes private property in all fields, but does not encourage, nor protect it) and, failing to provide clear definition of all properties in Romania, does not favor reform.

Reform of administration

Reform of the administration and of macroeconomic management started in 1990, but is far from being complete. The central plan and the quantitative quotas for supplies were abandoned after the first half of 1990, but electrical energy, water, natural gas and fuels were still centrally allocated at the end of 1993. Economic ministries continued to play a significant role in resource allocation and, not surprisingly the same institutions are called today to draft and execute 'restructuring' programs of different industrial sectors. After an initial attempt to reduce the number and powers of the economic ministries during 1990 (Laws 6/1990 and 46/1990), we have seen after the first half of 1992 a strong reinforcement of the old structures (even the re-establishment of some of the bodies that were virtually suffocating the Romanian economy before 1989). The number of Cabinet members grew from 17 in 1990 to 25 in late 1993. A law on local public administration and governmental provisions on local taxes were introduced (Law 69/1991 and Government Act 15/1992), but they do not provide proper financing, so administrative divisions still rely on allocations from the central budget. In spite of the efforts, including EU PHARE assistance to the Ministry of Finance, no proper budgeting techniques are used, and the state budget is prepared and executed mostly according to preexisting procedures. This is why, during all the transition period until now, the government failed to reach the normal deadline for submitting the budget act to Parliament, and worked in the first three or four months of the fiscal year with a provisional budget. In all cases, the debate on the budget became useless, as there was no other possibility but to validate the exercise. Because of the same facts, every year of the transition, Romania had had two budget acts, as a revision proved to be necessary (in all these years, inflation projections have been significantly lower than the real rates, and revenues seriously lower in real terms).[9]

Fiscal policy, tax reform and price decontrol

The most important tax policy measures, associated with the reform of administration and of prices, have been the introduction of the value added tax (Government Act 3/1992) - scheduled to enter into force on 1 January 1993 and finally introduced on 1 July 1993 - and of the profit tax for companies (Law 12/1991), measures that are supposed to be followed by the introduction of the personal total income tax and of withholding taxation of interest income.

Deregulation and price reform were thought to be key issues in Romanian reform and rapidly turned into one of the most important political debates. In the specific environment of the Romanian economy, with huge shortages and many products and services supplied by monopolistic structures, the gradual approach was estimated to provide a greater degree of acceptance and popular support and price deregulation occurred in several stages, from November 1990 to October 1993. Over this period of time, price controls have been removed, sometimes replaced with consumer subsidies and producer subsidies, that were, in their turn, removed, so that at the end of 1993 only few prices are still controlled (of course, one can argue that still the most important prices - energy, fuels, wheat, urban utilities - are controlled, but most of these products and services can also be obtained at uncontrolled prices, on the market, if the suppliers have no inputs from state resources). This gradual approach caused serious distress in the economy, as it produced several inflationary shocks and it provoked hoarding before the announced rounds of liberalization, demonstrating that: "... partial price reform is the worst of all choices".[10] Still in May 1992 managers of state-owned enterprises were complaining that: "... the state again interferes in detail in price formation, but without correlations and economic substantiation, bringing about great difficulties and big losses".[11] Wages still bear a certain level of control in the state sector (still the most important employer in the economy), through the use of a supplementary tax on the total wage fund, if the monthly level of this fund exceeds a sum that is substantiated by labor productivity in the company for the relevant month.[12]

Trade policy reform

Trade policy reform was among the fastest: in the second half of 1990, all restrictions on imports were already removed, the state monopoly on foreign trade activities abolished and a new customs tariff introduced. As a matter of fact, Romania, as a member country of GATT, as soon as the irrationality of the regime disappeared, started to fulfill its obligations as a Contracting Party, according to the Accession Protocol. Licensing of foreign trade operations was maintained for statistical purposes and only some quantitative restrictions on exports were introduced and maintained, with a view to balance the market of certain products in short supply, to prevent a drain of resources with still controlled prices or to observe trade agreements with some partners (the EU, for example). These measures preceded by more than one year the reform of the exchange rate mechanism, so, in an attempt to give some coherence and effectiveness to import liberalization, during the summer of 1990, some measures were introduced with respect to exchange controls: exporters were allowed to retain a percentage (variable according to ownership - state or private,

Romanian or foreign) of their export earnings in their own currency accounts. This legal provision, aiming both at stimulating exports and at providing a source of currency for imports, not coupled with proper measures in the banking sector, proved to be counterproductive and seriously contributed to the 'dollarization' of the Romanian economy.

Banking and financial sector reform

The reform of the banking and financial sector was among the latest in the sequencing of the Romanian transition. The Central Bank Act (Law no. 34/1991) and the Act on Banking Activity (Law 33/1991), introduced a two tier banking system, separating the two main functions of the former National Bank of Romania. Starting in the summer of 1991, the National Bank of Romania, as an independent central bank, followed a monetary policy aimed at bringing down inflation and enforcing financial discipline. The regulatory framework also entrusted banking supervision to the central bank and not to the Ministry of Finance. Four large universal banks, with a nationwide network of branches, appeared in this new market, as a result of the transformation. Other banking institutions have been licensed since by the National Bank of Romania, and about 20 banks are now legally constituted in Romania, including some foreign ones (Chemical Bank, Société Générale, Misr). Still, the number of banking institutions that are active all over the country is limited and clients have a limited choice, so there is no serious competitive pressure and banking is still rather monopolized. The same applies to insurance, although the former state-owned insurance company was divided into three, which are now competing with four other private companies that have been found and legally registered since the adoption of the Act on Insurance Activities. Interest rates were gradually decontrolled starting in 1992, following the rounds of price liberalization, and the central bank constantly increased its refinancing rate in an attempt to reach positive real interest rates. The introduction of a penalty refinancing interest rate of 260 per cent in November 1993 gave birth to many debates and protests in the Parliament. The first global compensation schemes for inter-enterprise arrears, as well as writing off most of the 'bad loans' that the banks inherited from the communist regime, were measures promoted by the National Bank of Romania which argued that a strong and healthy banking sector is a prerequisite for successful transformation of the economy. Even the recent move to place the largest loss-makers in the economy under special monitoring also originated in a study about the restructuring and the privatization of the banking sector.

The results of the implementation of the economic reform packages until now demonstrate that the Government still has a complicated apparatus at its disposal and a will to use it to avoid market forces. Analyzing the governance of the administration, a foreign consultancy company pointed out that powers and roles allocated to different governmental bodies are in conflict, that there is a lack of coordination, individual ministries and agencies work in black boxes and laws are not always observed.[13] As an example, one of the most flagrant breaches of regulations is the continuous process of dividing and merging state-controlled companies through Government Decisions. As an effect of the privatization regulations, since 1992 there is no 100 per cent state-owned company, all of them having the State Ownership Fund (SOF) and a Private Ownership Fund (POF) as shareholders for 70 and 30 per cent, respectively. As the State Ownership Fund is an independent institution and there is a shareholders' agreement with the Private Ownership Funds, any reorganization of a company would require a decision from both shareholders. There is no legal provision that could allow the Government to take decisions on behalf of the shareholders (not even on behalf of the SOF that represents the state), but the Romanian Cabinet does not even bother to inform the shareholders of such decisions.[14]

Industrial restructuring

Another specific problem of the Romanian economic reform is the large gap between the intentions of the Government to improve company management and efficiency, through the legislation and macroeconomic measures adopted, and the results of this policy at the microeconomic level. In this respect, it is possible that this has been an outcome of ignoring the differences between the role of changes in the institutional structure of the economy (or what is called by Daniel Bromley 'institutional transactions')[15] and the role of rules, procedures and regulations directly related to physical trading among different independent economic agents.

Management in the transition period

Duality between the change in environment and the inherited routines

The problem of company management was not ignored in the Romanian transition, but was not very well addressed either and too much was expected out of untrained managers acting in new conditions of decentralization, decontrol and exposure to market competition. The reform program has involved a massive upheaval in institutional structures, roles and processes in a very short period of time. The pace of change is so rapid that every week brings something new to the country's political, social or economic environment. On the surface, nothing today is the same as it was three years ago and next week something more will have changed. Underneath these eddying currents and new tributaries of change, much of the river still seems to be flowing along its old course.

To change its direction requires a shift in ways of thinking and behaving in management structures and hierarchies. Otherwise, previous behaviors will control the new management structures and processes and the old pattern of decision-making, which is inappropriate for a market economy, will continue.

Some examples

Decision-making and lack of decisiveness at lower levels. Examples of command economy behavior which still apply in Romania, as well as in other East European economies, are as follows:

1 all key decisions are sucked to the top, thus creating delay, bottlenecks and hassle;

2 managers at lower levels often operate as though they are waiting for the next instruction or law to be passed down, so that they can turn it into another instruction for transfer down to the next level where people are again waiting for the next command.

Drucker gives an accurate description of such persistence of old type behavior of these people "... that spent their lives as paper-pushers ...", the managers available in reforming countries:

> Nothing so much frightens people in former Communist countries, visitors report, as to be asked to make a decision. They are paralyzed by the fear of making a mistake. They hold endless meetings, call for more and more studies and in the end find a good reason why someone higher up should take the responsibility. Yet it is the essence of a market economy - and its strength - that decisions are made close to market and customer ...[16]

Such a description could not apply better than to the majority of managers from state sector companies and most of the central and local administration in Romania in 1993. The situation is furthermore aggravated by the unclear institutional framework, by corruption allegations and presumptions of guilt that prevail in our society.

All these manifestations continue despite the fact that in the framework of enterprise reform and privatization, the Romanian policy aim was to create an adequate market economy environment for producers and to accelerate and stimulate the creation of the private sector.

Commercialization, management and trade unions. The main aim of commercialization was to introduce a new system and management structure in the Romanian economy, by allowing a considerable degree of liberty in decision-making at the level of the company.[17] Comparing the corporatization process taking place in different Central and East European countries, Gabor Hunya pointed out that in the case of Poland (the process in this country is presented in detail in this volume by Jan Mujzel) and

Hungary, an important political task was in fact the dismantling of the over-powerful workers' councils, but "... in countries with formerly strict central planning, such Czechoslovakia and Romania, corporatization increases the influence and the responsibility of managers ..." which means that "... they become more powerful to resist centrally initiated programs (e.g privatization), but also left alone with inherited financial and technological burdens ...".[18] His assessments seem to be correct in the case of Romania, where management has not improved and both trade unions and the Government blame the managers of the still state-owned enterprises for most of the problems the economy is facing. In fact, many of the latter find it easier to remain under state control which still gives some guarantee for survival, supply with energy and imports or the possibility to use the state-owned company that they manage as a source of assets and firm contracts for their own newly established private company. This is how Romania reached a rather unique situation in that the trade unions are pushing for privatization (although they are conscious that this may be associated with lay-offs).

Corporatization as a global process was adopted in 1991 because of the evidence that decisions imposed by branch ministries for a whole sector (in terms of management appointments, capital expenditures and investments, allocation of physical and financial inputs and outputs) were virtually suffocating the enterprises. It was not only that many decisions were not perceived as correct at enterprise level, but also, due to the disappearance of the strong central planning authority, correlations between different sectors were completely ignored, with each ministry making its own policy (of course, aiming at what they were used to: ever increased physical output, high capacity utilization and full employment).

Such a system was causing a drain of resources that could not be supported by the Romanian economy. Transferring decision-making power and responsibility for managing physical and financial assets to the level of the companies was sought as a solution but, as a result of several factors it did not work well. The impact of the change seems to have been rather negative, with the state still unique or majority shareholder.

The influence of branch ministries. The transformations should have been accompanied by the dismantling of the former vertical integration in the so-called 'industrial centrals', but in practice the overly strong influence of branch ministries lead to the development of some very odd corporate structures. Afraid of losing control over the activities of economic units and the resource flows of the economy, branch ministries have imposed (using the idea that some integration has to be preserved for economies of scale that are resulting) the organization of several so-called 'holding companies'.

Such holdings were created, upon ministries' directives and approved by government resolution, as associations of commercial companies especially in the oil industry, metallurgy, heavy industry and agriculture. Holding companies are based on a cross-ownership structure, and are registered according to the company law (Law 31/1990).

We have witnessed in this case the development of artificially created structures that are neither real owners (as the real, ultimate owner is still the state), nor management companies, but which have decision-making powers and are continuing the inefficient allocation of resources and interference at individual companies' level, thus questioning the reality of changes in the organization of state enterprises. There are also such holdings that now have a powerful monopoly position, such as the artificially created Rafirom S.A., which developed from the former 'central' of the refining industry and will become (after the abolishment of quotas) the only domestic supplier of oil products.[19]

At the same time, even in the cases where 'holding companies' do not exist, the influences of branch ministries are strong, sometimes resulting in direct management. Also the branch ministries interfere with the attributions of the State Ownership Fund and the Private Ownership Funds in appointing the boards and the executives of the commercial companies, as well as in developing their own strategies and procedures for privatization.

Although under the existing legal framework there is no need for a ministerial approval for any decision taken at enterprise level and supplies can be obtained by direct negotiations with domestic or foreign suppliers, still many resources are (despite legal provisions) allocated in a preferred manner, and investment decisions (financed out of the companies' financial results) are especially subject to great interference.

There are several examples of orders issued by ministries to enterprises producing goods which are still in shortage not to sell their products to private enterprises or to sell them only to one state-owned distribution company. This has been recently the case with flour and wheat (by order of the Ministry of Agriculture and Food Industry, mills from the state sector were allowed to sell flour only to state-owned bakeries), with oil products, wood and even some manufactured products. Such discriminatory practices introduced by branch ministries refer also to customers from the private sector: while state sector enterprises are building huge stockpiles of arrears, sales towards private companies are allowed solely against cash. Another practice of the ministries that are still exercising their jurisdictions over most of the manufacturing plants in the country is to oblige companies to use only a state-owned export agent (sometimes created specially by the ministry).[20]

The recent adoption of a regulation in the field of management contracts was meant to clear these relations, but the measures so far adopted with a view to implement such a regulation show a tendency to use this act just to provide a more protected environment for the managers of state-controlled units, closely connected with the bureaucrats in branch ministries, with the majority party (which also controls the board of the State Ownership Fund), in an informal network of economic, social and political interests.

Conclusion

In such a context, it seems that for Romania it is quite clear that the very necessary restructuring of the economy cannot take place before privatization, in the absence of an efficient principal-agent relation. The restructuring is difficult at present because - as Stavros Thomadakis emphasizes in the last chapter of this volume - the transitory character of the mixed economy determines a short-term orientation of the managers of the state-controlled companies. In many cases they are more preoccupied by how to pay wages than how to build the future of the company. Even representatives of one of the most conservative government bodies, which is the Ministry of Industries, estimating the necessary investment effort for restructuring industrial sector in the next four to five years to $ 14-16 billion, now think that:

> ... privatization is imperative also in the context of this tremendous investment effort. The new owners will be the most capable to make the restructuring process at microeconomic level efficient. The delay of privatization in the industrial sector may be considered, as a consequence, as an obstacle also for the restructuring.[21]

Privatization

The concept of 'privatization' used in this paper refers to the organized ownership transfer from the state sector in favor of private owners, and not the overall system of transformations that have as a result the increase of the share of the private sector in the economy. Derivative forms of private use of public assets (such as lease or management contracts) are not covered in this section, nor is the setting up of new private enterprises by domestic or foreign investors. This differentiation was present from the beginning of the reforms in the debate among government experts, consultants and scholars, who referred to the second category as 'private sector development', that is all measures and activities that lead to the development of the private sector without implying a transfer of ownership titles.

In Romania, a complex privatization scheme was developed, starting from the need to address specific features among which the most important were:

1 the corporatization process of former state economic units;

2 the lack of a significant amount of domestic savings coupled with the huge asset value of the state sector;

3   the political promise to give an equal share to every citizen for participation in the privatization process;

4   the lack of development of the banking and financial sector;

5   limited knowledge about the functioning of the market and free enterprise system.

This scheme was included in three main pieces of legislation: Law 15/1990 on the commercialization of state enterprises, Law 58/1991 on the privatization of commercial companies and Law 114/1992 on the by-laws of the Private Ownership Funds, the privatization strategy, the regime of the Certificates of Ownership and the rights of holders of such certificates. The regulatory framework is furthermore detailed by numerous government resolutions, decisions and orders of different ministries, as well as by common norms and procedures of the main institutions involved. These are:

1   the National Agency for Privatization (NAP), the specialized body of the Government, assigned by law 'the coordination, guidance and control of the entire privatization process' (art. 63 in Law 58/1991);

2   the State Ownership Fund (SOF), an independent public institution that holds 70 per cent of the shares of all former state-owned enterprises and must divest its portfolio at a rate of at least 10 per cent a year, as well as provide the financing needed for privatization and for preparation of the companies for privatization (including their restructuring);

3   the Private Ownership Funds (POF). There are five POFs, public companies that hold 30 per cent of the shares of the former state-owned enterprises and are the issuers of Certificates of Ownership as bearer securities and have as their main goals to restructure their portfolios so that they reach a stage in which they can turn into closed-end mutual funds, to maximize the value of the certificates and to provide to the holders of certificates brokerage services for the exchange of certificates against shares in companies that are being privatized.

Typical features of the Romanian privatization scheme

There are some features that individualize the Romanian privatization scheme and legislation among other privatization schemes in the region (compared with privatization programs in Poland, Hungary and the Czech Republic presented in other chapters of this volume):

1   The free distribution plan. According to an early act (Law 15/1990), when no privatization plan was yet drafted, 30 per cent of the equity in state-owned enterprises was given, free of charge, to all adult Romanians. This legal provision raised a vivid public debate, but while specialists were more concerned with the effects of such a measure, the debate concentrated more on the equity and ethical issues, thus questioning only the size of the share that was supposed to be freely distributed. Later, with the enactment of Law 58/1991, a complicated system was introduced to solve the give-away scheme. The solution resulted in the distribution of 15.5 million bearer securities that are neither privatization vouchers nor proper shares and that have been constantly traded on an unorganized market at about ten per cent of their nominal value.

2   The 'specialization' principle. The success of the privatization scheme relies upon effective action of the main actors (SOF and POFs, under the coordination of NAP) that are supposed to perform the tasks entrusted to them by the particular regulations, tasks that are also determined by economic interests. One major problem, identified from the very beginning as a potential threat, is the nature of the SOF: an institution, organized as a company, that is supposed to dismantle itself. Also, the debate on the law pointed out that there is a real danger for further development of the SOF as a super-ministry, a bureaucratic monster, that will suffocate the economy even more than the former State Central Planning Committee. Strong governance and supervision of the process by the specialized government body (the NAP) was necessary to avoid such dangers, as well as to provide control to the boards of the POFs, in the absence of proper shareholder control.

3   The object of privatization. The Romanian privatization plan comprises solely industrial and agricultural properties that were previously organized as commercial companies (that is public joint-stock companies (corporations) registered under the common company law with the Trade Register). Note that mainly as a result of the reorganization of state administration, additional commercial companies are formed which prolongs the specialization process and keeps certain industries under state control.

4   The separation of financial flows from the state budget. Under the law, both dividends paid by companies and proceeds from privatization are going to the SOF and POFs, and not to the state budget. Such a pattern could also be derived from the quality of these institutions as shareholders, under the common law, but the Privatization Act also has specific provisions to reinforce it, as well as provisions as to the destination of funds. The proceeds obtained by the SOF have to be used to finance the privatization process (consultants, logistics, financial

rehabilitation of companies to be privatized, extension of credit to buyers, etc.), while the POFs have to properly invest their revenues so that the market value of the Certificates of Ownership (in fact, their shares) can be maximized. Revenue out of sales of companies' excess assets are retained by the companies.

5   The privatization of units as independently functioning entities. Due to the corporatization process that preceded privatization, in Romania there was no room for direct sale of independent, individual state-owned assets, as all were incorporated into a company. The law stipulates a so-called sale of assets program, but the object of sale is defined as 'assets belonging to a company that are organized so that they make, or can make independently functioning units'. Such units are publicly auctioned, the decision to sell is at the level of the company and the revenue goes, naturally, also to the company. The rest of the privatizations are sales of companies or of a significant package of shares in a company, by the SOF or by the POFs (table 6.1).

6   The 'Pilot' Program. As it was clear from the very beginning that the setting up of the specialized institutions was going to take time, the Privatization Act enacted the possibility that a special program of 'early privatization' would be executed by the specialized government agency, with a view to experiment the feasibility of the scheme, to provide some initial cash revenue for the SOF and for the POFs and to give some momentum to the privatization process by showing the public the example of some successful privatizations. Training of the necessary manpower for the 'big privatization' was also considered.

Another special feature of the Romanian privatization scheme is that there was very little consideration of private initiative. According to the program, the citizens and institutionalized private investors have a rather passive role, although their active participation is allowed. There is, however, no stimulus of such activity, nor any component of the system left solely to private initiative. Such an institutional approach encompasses a great risk that the accidental failure of an institution, which may be caused by bad staffing or the nomination of a weak (if not adverse to real privatization) board or head, could result in the collapse of the entire system.

Table 6.1
NAP privatization results

	Dec. 1992	Nov. 1993
Completed transactions (100 % shares sold) for sale of shares, out of 32 proposed companies (in the Pilot Privatization Program)	15	22
Proposed transactions for sale of assets (in the Sale of Assets Program)	6,105	7,602
in trade	2,976	3,509
in tourism	1,711	2,114
in industry	526	623
Employees	41,268	50,112
in trade	5,942	7,143
in tourism	11,890	13,949
in industry	12,974	16,125
Completed transactions for sale of assets (in the Sale of Assets Program)	1,514	2,847
in trade and tourism	1,259	2,290
Employees	7,967	13,218
Distributed Certificates of Ownership	15.54 mln.	-

Source: NAP, 1993.

For the purpose of reducing employees' resistance to privatization, special facilities were included in the Privatization Act, such as a discount on the offer price and credit at preferential terms. During the execution of the pilot program, the interest showed by employees and managers, on one hand, and the limited demand from the part of other investors, on the other hand, have brought the relevant decision-makers to put forward a standard

Management/Employee Buy Out scheme as a solution for smaller companies. Such schemes proved feasible and they were introduced by special norms, jointly signed by the NAP, the SOF and the five POFs, with a view to accelerate privatization. There are, however, hazards associated with large-scale use of this method taking into account the cases of suboptimal decisions (documented by Bilsen in this volume) and other limits (described by Giday and Sári-Simkó for Hungary or identified by the evaluation report of the NAP in Romania).

As far as methods that are allowed to be used in the privatizations are concerned, the existing legal framework gives room for almost all known forms, from public flotations to trade sales or even privatization through liquidation.

Achievements by the end of 1993

The Pilot Privatization Program started in late 1991 and ended in August 1993 (for the results see table 6.2). A report of the National Agency for Privatization stated the following:[22]

> Until now, the privatization process evolved excessively slowly, a situation illustrated by the fact that:
> 1   within the framework of the pilot privatization program, only 22 companies have been privatized;
> 2   according to the standard Management/Employee Buy-Out procedures for privatization of smaller companies only 188 companies out of 3,784 have been privatized;
> 3   the State Ownership Fund privatized through trade sales (negotiated sales) 14 medium-sized companies, representing 0.39 per cent of the total equity of medium sized companies, and 5 small companies;
> 4   according to the law, 2,849 assets have been publicly auctioned, as compared to an initial number of 6,097 assets published for sale.

To such remarks of the specialized government agency we can add one of an outside observer, who says: "Romania's privatization program is in a horrible mess".[23] That may sound too critical, but truly reflects a difficult situation of one of the key components of the transition. This complex situation is furthermore complicated by the phenomena of 'spontaneous privatization' that may leave too little to privatize according to the legal framework when the relevant bodies would consider taking some more energetic action.

Table 6.2
Completed transactions in the Pilot Privatization Program

Name	Equity HGR 945/1990 (1,000 lei)	Industry	Selling price (1,000 lei)*	Number of employees	Future investments (1,000 lei)*
Ursus Cluj	212,720	Brewery	320,232 + DM 1.22 mln.	472	DM 2 mln. (in 3 years)
Arta Grafica Bucuresti	123,285	Printing	123,285	460	-
IPCT Bucuresti	27,305	Design for building & construction	65,532	350	-
Vranco Focsani	234,335	Clothing	71,624 + $ 489,822	4,126	$ 5 mln. (in 5 years)
Agroind Focsani	2,551,156	Agricultural products	498,490	288	150,000 (in 3 years)
Conimpuls Bacau	34,984	Building & construction	47,250	538	100,000 (in 3 years)

Table 6.2 (continued)
Completed transactions in the Pilot Privatization Program

Name	Equity HGR 945/1990 (1,000 lei)	Industry	Selling price (1,000 lei)*	Number of employees	Future investments (1,000 lei)*
Unicom Galati	198,075	Building & construction	220,416	1,538	50,000 (in 3 years)
Itels Iasi	97,540	Building & construction	110,000	308	30,000 (in 3 years)
Eurovitis Urechesti Vrancea	283,930	Agricultural products	157,887	26	3,000 (in 3 years)
Comrepi Borzesti	38,700	Building & construction	119,970	370	30,000 (in 3 years)
Alfa Braila	15,430	Building & construction	44,576	326	30,000 (in 3 years)
Unicon Bucuresti	5,862	Building & construction	28,152	362	150,000 (in 5 years)

163

Table 6.2 (continued)
Completed transactions in the Pilot Privatization Program

Name	Equity HGR 945/1990 (1,000 lei)	Industry	Selling price (1,000 lei)*	Number of employees	Future invest-ments (1,000 lei)*
Electric grup Iasi	137,410	Building & construction	137,410	492	30,000 (in 2 years)
Industria Carnii Pascani	89,732	Meat processing	99,959	93	50,000 (in 5 years)
Magura Codlea	957,990	Furniture	724,472	1,700	$ 4 mln. (in 5 years)
Terra S.A. Bucuresti	116,200	Foreign trade	450,021	72	-
Fruleg Alexandria	2,065,345	Vegetable processing	$ 2.6 mln.	579	$ 78 mln. (in 18 months)
Confectia Tîrgu Secuiesc	508,100	Clothing	850,000	2,337	1,005,000 (in 3 years)

Table 6.2 (continued)
Completed transactions in the Pilot Privatization Program

Name	Equity HGR 945/1990 (1,000 lei)	Industry	Selling price (1,000 lei)*	Number of employees	Future invest- ments (1,000 lei)*
Rudbin Rudeni Chitila	95,540	Furniture	143,310	189	-
Milcov Focsani	570,225	Clothing	76,981 + DM 1.7 mln.	937	DM 10 mln. (in 5 years)
Electrofar Bucuresti	789,215	Lighting	844,460	1,106	$ 6 mln. (in 4 years)
Apulum Alba Iulia	3,953,347	Porcelain	3,992,884	2,498	1,400,000
Total	13,106,462	-	9,706 349 + $ 3.1 mln. + DM 2.92 mln.	18,591	3,028,000 + $ 93 mln. + DM 12 mln.

* unless noted otherwise
Note: The deadline for the Pilot Privatization Program was August 10, 1993.
Source: NAP, 1993

Through the process of setting up joint ventures or increasing the capital by issuing stock distributed to management boards and employees, a large part of the state property escaped the system and was transferred towards private owners (Romanian or foreign), without proper evaluation and no competitive procedures. Of course, one might argue that privatization being the target, this is a good effect,[24] but, under the circumstances, such a 'natural' privatization (based on preferred treatment) did not always result in a proper allocation of resources taken away from the public sector. Commonly referred to as nomenklatura privatization (as in many cases such actions were undertaken by directors and executives who preserved top management positions from the old regime), this process is seriously resented by the population and creates a distorted image about the reforms.

The State Ownership Fund and the Private Ownership Funds were set up in late 1992 and became operational at the beginning of 1993. Certificates of Ownership were distributed free of charge, except for a small fee, which was meant to cover the distribution costs. Each citizen received a Certificates of Ownership Booklet which contains five certificates, one for each POF. The distribution process finished by the end of 1992, the first nominal value was announced at the beginning of April 1993, and revised quarterly since, according to the law (table 6.3).

Table 6.3
Value of the Private Ownership Funds' Certificates of Ownership

POFs	Value between 20 April- 29 June 1993	Value between 29 June 25 Sep. 1993	Value after 25 Sep. 1993
Banat-Crisana Arad	23,000	25,000	28,000
Moldova Bacau	29,000	30,000	30,000
Transilvania Brasov	29,000	30,000	30,000
Muntenia Bucuresti	27,000	28,000	29,000
Oltenia Craiova	27,000	28,000	28,000
Total	135,000	141,000	145,000

Source: NAP, 1993.

Main obstacles and problems

The main obstacles and problems that are causing delay to the official scheme of privatization in Romania, could be summarized as follows:

Problems associated with the functioning of the existing institutional arrangement

Unclear institutional arrangement. In the privatization process too many institutions are involved in decisions - the Cabinet, the Parliament,[25] the NAP, the SOF, the five POFs, the economic (branch) ministries and the local administration, and in the absence of a clear delimitation of competencies and powers, contradictions and overlaps develop. One of the main causes lies in the fact that the Privatization Act did not explicitly stipulate that administrative bodies other then the ones described in the law should observe the measures taken by the specialized institutions.

Legal status of the State Ownership Fund. According to the Privatization Act, the SOF is a public institution of general interest, with commercial and financial activity that is responsible to Parliament. As a general rule, a public institution like the SOF is established by special (organic) law - the rule that was observed in this case - and should be subject to control from the part of the specialized body of the executive. The functioning of the SOF under the control of the Cabinet is also requested by the responsibility of the executive for an effective action for reform, of which privatization is a main component. Until now, in practice, the SOF assumed the status of an independent institution, solely responsible to Parliament and its relations with this body have been limited just to some discussions in the specialized committees. On the grounds of such interpretations, the SOF did not cooperate with the different bodies of the executive in the field of the coordination and unitary execution of the privatization and restructuring processes, thus endangering the coherence of the macromanagement policy of the Cabinet.

Non-observance of some specific legal provisions. There are two distinct forms that occurred in the process of implementing the specific privatization regulation:

1 the non-observance of some important legal provisions by the institutions involved in the privatization process, as there were no penalties enacted regarding such violations; and

2 delayed implementation of some measures and contradictory actions by different bodies of the administration.

From the first category, the most damaging are those that affect the governance of the privatization process (which relies almost exclusively on administrative action). For instance, none of the provisions in the Privatization Act calling for annual reports and other information have been observed in more than one year and a half by the SOF.

The second category of implicit infringements, by long delayed measures, contributes to the confusion, especially affecting corporate governance of still state-controlled enterprises and decision-making in the privatization process as well as investors' confidence. The most significant among these are the delays in the issuance of title deeds attesting land ownership (both for individuals and for companies), in the transfer of the 'transfer' titles of the shares in 'commercial companies'[26] and in the appointment by the SOF and POFs of their representatives for the shareholders' assemblies in the commercial companies. The delays in the distribution of the 'acts of transfer' allows many abuses by company managers, branch ministries, local administration and even the Cabinet. Some examples illustrate very well the extent of confusion that is thus generated.[27]

Internal contradiction caused by the dual nature of the State Ownership Fund. The one and a half year of activity of the SOF brought to the surface a major contradiction between the position of the SOF as an administrator of the equity still held by the state in the 'commercial companies' and the legal mandate of the SOF to divest its portfolio at an accelerated rate. This contradiction is demonstrated by the trend that can be traced in the activity of the SOF to preserve the most profitable shares and to try to get the highest cash value for the shares that it is offering, so that it can generate enough resources for restructuring and even subsidizing the activity of large state-controlled units. Furthermore, as the staffing of the SOF is far from being satisfactory, this institution cannot perform the tasks of administration of this huge portfolio and of divesting it at the same time.[28]

Activity of the Private Ownership Funds. The five POFs started to behave as mutual investment funds although their transformation into regular financial intermediaries is allowed only after a five year term (Art. 14 in Law 58/1991) and only when they reach a stage at which the majority of the Certificates of Ownership have been exchanged against common stock of companies. They are delaying the offering of the shares that they hold against Certificates of Ownership and tend to monopolize the secondary trading of these certificates. The investments made by the funds (and, generally, their entire budget execution) are carried out in the absence of proper accounting norms and of the possibility of an effective control by the shareholders, thus allowing excessive expenditures or highly risky placements.

Evaluation and pricing. As there are no markets for productive assets and for enterprises and no organized secondary trading of shares, valuation of enterprises and of equity poses great difficulties to the privatization process. In the absence of markets and due to the specific administrative approach to privatization, in Romania pricing of the shares offered in the privatization process is based on the 'net accounting value' of the firm, that is the inflation adjusted historical cost of the assets of a particular company. This approach, which is in no way related to the future stream of income that a company can generate, is of no significance for most of the prospective buyers and maintains the inflated level of the minimal price accepted. As on the specific privatization market in Romania we have to deal with an excess supply (if we compare the total book value of the shares and the assets involved in the 'seven year plan' with the amount of capital available); such a rigid attitude further reduces the chances of success.[29] Ignoring the fact that the final transaction price is the result of negotiations between the buyers and the offerer, several institutions of the administration have contested several transactions that were concluded in the 'Pilot' Privatization Program. They accused the National Agency for Privatization of 'selling state assets below their real value', asking for an administrative breaking of the contracts (practically, a re-nationalization of the companies) and prohibiting sales. What is not understood is that the valuation of a company is different from the valuation of tangible assets or stock and that there should be no administratively set relation between the valuation (which can reach a wide variety of estimated prices, according to the method used) and the actual price in a privatization contract (which is the result of a transaction). There is a great amount of confusion with respect to categories like equity and capital, assets, nominal value and market value of securities and all these, associated with accounting practices not properly reformed, are due to the widespread lack of knowledge about how the market system functions. They are reducing the various methods and techniques described in the privatization regulation to simple words.

Definition of property rights. As the situation of property rights is also described in section three, we refer here only to some of the effects that are the most damaging to the privatization process:

1   the majority of companies are not in possession of a legal document that would certify the extent of their rights to the land which they currently posses (ownership, long-term lease, rent);

2   there is no clear delimitation between public properties and the private domain of the state and of the municipalities and communes;

3   provisions of Article 77 in Law 58/1991 (regarding compensations for
     properties acquired by the state through abuses) have not found a
     reasonable legal settlement and nothing has been enacted with respect
     to restitution of nationalized properties, while claims of former owners
     who can document their actions are taken into account by courts of
     law;

4   different interpretations of the rights of companies that are controlled
     by foreigners to own real estate properties are used by local and central
     administration bodies according to specific interests.

Tax regime. No specific tax incentives for privatization are in place in
Romania and until now this issue was not even considered. More than that,
the Romanian tax regime discriminates between domestic and foreign
investors (foreign investors are given a more preferable tax treatment than
domestic investors) and, what is worse, between newly created enterprises
(which are allowed tax holidays from profit taxation for up to five years)
and privatized enterprises (which are subject to taxation). Especially the
latter is counterproductive for privatization because it orients the available
capital towards the creation of new companies instead of favoring the
acquisition of existing stock.

Equal treatment of economic agents and access to resources. Many
discriminatory practices were identified by several surveys of the private
sector and of the enterprises privatized within the pilot privatization
program:

1   The banks have different operational norms, according to the forms of
     ownership and privatization caused in all cases serious delays in
     banking operations for the privatized companies, as they had to rewrite
     and renegotiate all their arrangements with the banks (this is the case
     with state-controlled banks, but they accumulate most of the banking
     operations).

2   Treatment by the administration is much more favorable for companies
     with state participation (licensing, legal registration, etc.).

3   Activities of control and prevention of economic crimes and tax evasion
     have concentrated on private sector companies.

4   Access to information, logistical support in marketing, foreign trade,
     research and development, foreign technical assistance, and training is
     easier for companies with state participation.

5   Access to material and financial resources and to government contracts
     is often restricted to private sector companies (especially if there is no
     personal connection with an individual that is influential in a ministry

or a large state-controlled unit, as such restrictions are just rarely taking the form of publicly disclosed regulations).

Strategic objectives. The privatizations concluded until now reflect a non-systemic approach to the process, chaotic actions that took place more likely because of particular personal interests than as a result of an organized offering. With the exception of the Pilot Program (under which the clear objective of achieving successful privatizations while experimenting with the main techniques enabled the NAP to fulfill most of the goals of the program), there was no attempt to define or to execute specific programs that could benefit from the synergy provided by the concentration of resources.

All these problems, adding to the shortage of capital and to a constantly negative attitude towards privatization promoted by the media made the comprehensive legal framework specific for privatization rather useless and contributed to the aggravation of the economic downturn in Romania. In the absence of a rapid reassignment of ownership over the productive assets (with real, identifiable owners taking over 'nobody's property') the structural adjustment made little (if any) progress. Resources for restructuring were diverted towards consumption, even external financing sources obtained under rather strict conditionalities were abused, the drain of resources continued and this is how, after three and a half years of reforms and two macrostabilization packages, inflation reached its peak, there is no financial discipline and imbalances are growing.

Reform of the financial sector

A move to a market-based economy must involve the financial sector at an early stage (the chapter based on the Bulgarian experience contributed by Panov argues this assertion). Market forces cannot determine the price volume of output properly if one of the most crucial factors of production - capital - is priced artificially and arbitrarily.

In the case of transition economies, where the problem of promoting development policies is coupled with the necessity of introduction of new institutions, large-scale ownership transfer and massive learning, the role of banking and financial institutions, as well as that of financial markets for promoting a learning-by-doing approach is indeed much greater.

Financial sector reform has been identified as one of the key elements of success in the case of development of a market system and yet, in Romania, as is also the case for other previously communist countries, there is limited progress.

The main measures that were introduced concentrated on the banking institutions and activities, so that one could expect an evolution in the direction of a 'German-type' economy, with a strong, healthy banking sector that closely monitors the enterprises and exerts a considerable influence on the management of companies. These measures were: a two tier banking system with a relatively independent central bank, a legal framework for banking activities allowing new entries into the market, interest rate liberalization and a certain degree of financial discipline for the state sector, more responsible monetary and credit policies, and a market determined exchange rate. An important measure was to provide a 'clean' start for both banks and enterprises and in this respect in 1990 most of the non-performant loans carried on the banks' balance sheets were wiped off (the budget was still able to cover such losses originating from the arbitrary allocation of capital during the centrally planned economy). Unfortunately, ensuring a clean start was not everything. In response to the restrictions on growth in the money supply, state-owned enterprises reacted by not settling inter-enterprise debt, thereby exacerbating the effects of the restrictive monetary policy and aggravating the liquidity shortage. In theory, a blockage of this kind can be relieved by compensation - the 'netting out' of inter-enterprise debt through a clearing house operated by an entity such as a central bank. The success of any such scheme turns on the capacity of all the participants to meet their obligations. Only the National Bank of Romania was able to act as a clearing house for the very high demands of the two schemes introduced in 1990 and 1991; more serious was the fact that some of the larger state enterprises, with liabilities under the scheme greatly exceeding their receivables, could not meet their commitments. This latter situation led to chronic payment arrears in the Romanian economy. Arrears build every year to enormous amounts, making useless the efforts of the central bank and the Government to enforce financial discipline and to properly control the expansion of the money supply. Faced with a difficult choice between closing down the responsible (and, in most cases, non-viable) enterprises - with the risk of social unrest and political adversities - and providing such enterprises with resources - with the risk of aggravating the inflationary pressures, the Government first chose the latter course. A third compensation scheme, originating in August 1991, should have ended in April 1992, but it was never completed and arrears continued to build. A study commissioned to a team of Western consultants that investigated in-depth, from the banks' perspective, this issue, came to the conclusion that approximately 100 enterprises - the largest loss-makers in the economy - were responsible for the continuous generation of payments arrears and the proposal was to isolate these companies and their banking operations into a separate financial flow, until a solution is found about what to do with them. The idea, which was welcomed by the Government and by the IBRD, resulted in the setting up of a Restructuring Agency (December 1993). Another measure, more likely to produce effects, was the introduction by a special act (Law 76/1992) of special provisions concerning the insolvency of state-controlled units, followed in July 1993 by a Government decision about the institution of a 'special financial monitoring regime' for insolvent

172

companies and the reaffirmation of the validity of the bankruptcy procedures in the old Commercial Code.

A lot was expected from the banks, which were supposed to induce the necessary financial discipline and responsible behavior at enterprise level. The development of short-term funds (money markets) was needed for a better financing of the real economy, the foreign exchange market was supposed to improve allocation of scarce 'hard' currency and the stock market was very much wanted for speeding up privatization, for improving company governance and performance, as well as for supporting the free distribution scheme of securities issued for privatization.

Until now, we have seen a reform of the banking system consisting of the introduction of two main pieces of legislation (the central bank (National Bank) Act and the Law on Banking Activity), and a limited foreign exchange market mechanism. There has been no privatization of a banking institution,[30] there is still no securities market, no inter-bank market for short-term funds and a lot of financial resources are wasted because of the absence of a proper framework for channelling them for investment purposes.

Several private banks have developed, but the market is still largely dominated by the state-owned banks. There is only one savings institution (CEC - the pre-existing savings bank) and the central bank (which is responsible for banking supervision) is not in favor of dismantling this monopoly.

The savings bank (CEC) is the main, if not unique, supplier of funds for re-financing the existing network of commercial banks. Sizable surpluses recorded by the State Social Insurance Fund, or by the Fund for Unemployment Indemnity both in 1991 and 1992 were kept in accounts with the central bank, with an interest about five times lower than the interest on the market. On the other hand, the deficit of the state budget was financed through interest-free loans from the central bank (a rather inflationary practice).

Due to the approach of the National Bank and to the influential lobby of the managers of the state-owned banks, huge organizations have been preserved, to the purpose of conducting a 'universal banking' type of business, which results in the banks continuing to work very much as they did during the previous regime. Current plans and privatization strategies do not involve any of the major banking institutions, nor their restructuring, but their auditing and eventual recapitalization. The state insurance organization was split into three insurance companies that are totally separated from the rest of the financial institutions, although they might have available liquidity.

The issues related to the securities markets are quite complicated, because of the poor communication between several institutions involved: Ministry of Finance, National Bank, Private Ownership Funds, National

Agency for Privatization, State Ownership Fund, and commercial companies.

Everybody is awaiting the law on securities trading and the stock exchange which was not yet passed and in the meantime, many securities have been issued according to provisions of different pieces of legislation and unorganized markets are developing. According to a rather odd decision of the Cabinet, in 1991 the preparation of the draft bill on securities and the stock exchange was assigned to the central bank, which is also concerned with developing an organized secondary market. A first draft submitted during 1992 was not found satisfactory, so the National Bank, assisted by a group of Western advisors, has prepared a new one, not yet submitted to the parliament.

On the other hand, in September 1992, by government resolution, a National Agency for Securities was created, subordinate to the Ministry of Finance, and it issued specific regulations through orders of the Minister of Finance. At the end of August 1993, a special ordinance of the Cabinet changed the status of this agency by giving it more powers to issue and enforce regulation in the field of securities trading, investors' protection and financial intermediaries, while a second piece of regulation enacted measures related to securities trading in the absence of the stock exchange.

The stock exchange is not viewed as an association of traders, but as a public institution, organized and regulated by the public authorities. Certificates of Ownership, shares of several privatized companies, as well as shares and bonds issued by newly established private companies, are traded in the streets. Shares of the company Ursus, the first privatization through public offering of shares, are selling more than ten times above the par value, and other private securities are also selling fairly well, substantiating the idea that the stock-market might prove to be successful.

The Government issued treasury bonds last year, but the issue was undersubscribed. The Treasury department in the Ministry of Finance continued to issue treasury bonds (maturing in six months, with a fixed interest of 50 per cent per year, in three different values of lei 50,000, 100,000 and 200,000), but faced the same lack of success.

No doubt that it was so, as, by an extremely odd decision, the terms of the issuance prohibit the participation of institutional investors (the order of the Ministry of Finance explicitly states that "... treasury bonds are only sold to physical persons through CEC counters ...").

In the case of the Certificates of Ownership, the market value in the streets is about one tenth of the nominal value announced by the Private Ownership Funds, which should, according to the law, take action for sustaining the 'market value of the certificates', together with the specialized government body (the Privatization Agency). This is why there was an attempt to organize a secondary market together with the Romanian Bank for Foreign Trade, which was done (neglecting both regulations of the National Bank and the Finance Ministry), but the volume of the transactions is extremely thin. Practically nothing exists, in terms of resources or mechanisms, to prevent a speculative bubble with Certificates of Ownership.

For 1994, the five POFs are preparing a concerted offer of shares in about 50 companies against Certificates of Ownership and a common position with respect to the secondary trading of these certificates.

The best foreseeable prospects for developing a capital market might be only the result of the activity of an entrepreneur. A core of a future market might develop due to the specific interests of the five Private Ownership Funds, which are 30 per cent shareholders in about 7,000 companies. Such a capital market could, however, play a limited role in exercising enough pressure on companies' management, and the same applies to the POFs as intermediaries. As for other potential intermediaries, it should be mentioned that although the existing legislation now allows other entries, there is no practical possibility that such intermediaries could acquire majority interests in companies, so that they are able to play a significant role in company management. In the meantime, the banking institutions are far too weak to provide a mechanism similar to the German model (where banks play a major role in enforcing financial discipline on enterprises), so management behavior in Romania is very likely to continue to be influenced more by the ministries and the State Ownership Fund (with a strong pattern of old regime practices).

Concluding remarks

The ultimate aim of reforms is to increase the satisfaction and living standards of the largest part of the population and/or of the most significant interest groups. The success of reforms is determined, among others, by the quality of the measures designed, their sequencing, effectiveness in introducing the changes and the attitudes of potential pressure groups that develop from different individual expectations about gains/losses from the changes. If sizeable reform packages are introduced over too long a time period, the interplay between potential losers and winners, under a democratic system,[31] may result in non-effectiveness of the packages or even in adverse results. Reforms can be much more effective (in terms of implementation and not necessarily in economic efficiency gain terms) under totalitarian regimes, when those that are hurt cannot oppose the changes in any other but violent ways.

In Romania, the interaction between old structures at the intermediate level associated with some prominent reformed communists that grabbed key decision-making positions and the forces for change produced a stagnation of the reform process, after the initial momentum. Old structures have been successful in delaying privatization and, without large-scale privatization, the whole transition to the market system is endangered. Without privatization, development of markets for productive assets is extremely slow and marginal, while "... the existence of markets for productive assets is the most important feature of a market exchange system based on private property, capitalism".[32] One serious problem of the Romanian transition is that it has failed to introduce a clear definition of ownership rights and of contracts. The concepts implied by neoclassical economics are:

> Rational individuals will compete not only to maximize their utility within a given set of rules, but also seek to change the rules and achieve more favorable outcomes than was possible under the old regime. A reduction in the cost of seeking changes in the structure of property rights will upset an existing equilibrium and lead to a new set of rules and a new distribution of wealth. Theoretically, only one set of rules will maximize the wealth of a nation.[33]

These concepts have their application in the example of the transition, but further investigation into the role of property rights in the transition is indeed necessary.

Transition packages have not improved company management in Romania, although such an effect was deliberately sought, especially during the last year. Imperfections of the legal framework, interference with political activity, lack of knowledge and of skills and inconsistencies of the theoretical substantiation of the reform measures are delaying progress in this direction, which is a necessary prerequisite for the improvement of resource allocation through decisions at the microeconomic level.

Improvement of company management in transition economies is not going to come too soon. The process of undoing 40 years (or more) of incorrect values, wrong incentives and poor policies is going to be a long one. As shown especially by the experience of Hungary and Poland, the development of capital markets can play a role in this process, but a limited one.

However, it is worth mentioning that an indirect factor that can substantially contribute to the improvement of company management in Romania is the development of the private sector as such. The contribution of this factor is twofold: on the one hand, the development of the private sector is a 'management school' for more and more people; on the other hand, the activity and economic results of the private sector influence the functioning of the other actors in the economy.

The potential of this sector is shown by the fact that at the end of 1993, after only two and a half years, the private sector represented in Romania about 26 per cent of GDP, about 30 per cent of foreign trade, and almost 50 per cent of the service sector.

As regards the development of new stock markets, the evolutions of the last decade bring strong evidence in favor of the role of capital markets in economic growth, by facilitating the process of introducing a market system.

Free, market-driven financial markets are essential to determine the price of capital realistically on the basis of return expectations of investors and to ensure that such capital is used efficiently. Competitive financial markets transmit efficiency to the real sectors of the economy.

Table 6.4
Statistical data and breakdown by size of state-owned commercial companies to be privatized

	Small	Medium	Large	Total
Number	2,677	2,510	675	5,862
% of total	45.67	42.82	11.51	100.00
Equity capital				
(bn. lei)	397,0	2,903.9	8,359.4	11,660.3
% of total	3.40	24.90	71.69	100.00
Turnover (bn. lei)	1,400.7	3,738.7	4,596.3	9,735,7
% of total	14.39	38.40	47.21	100.00
Revenues (bn. lei)	1,373.9	3,551.1	3,884.8	8,809.8
% of total	15.60	40.31	44.10	100.00
Pre-tax profit (bn. lei)	46.6	165.2	199.7	411.6
% of total	11.33	40.15	48.52	100.00
Employees	450,665	2,218,622	1,493,264	4,162,551
% of total	10.83	53.30	35.87	100.00

Small CCs equity capital $< =$ lei 400,000,000
Medium CCs equity capital $>$ lei 400,000,000 and $< =$ lei 3,000,000,000
Large CCs equity capital $>$ lei 3,000,000,000
Note: only data on companies with state participation for which financial statements are available and identifiable based on the Ministry of Finance's fiscal code
Source: NAP's database, 1993.

Table 6.5
Small-size companies privatized by the SOF and POFs (completed transactions - 100 shares sold as for 22 November 1993)

No. of companies	Equity	Proceeds for SOFs	Proceeds for POFs	Employees
156	lei 24.4 bn.	lei 31.3 bn.	Certificates of Ownership	33,531

Source: NAP, 1993.

Economic growth requires that new and growing companies expand their capital bases. A large, liquid stock market facilitates the process, giving the owners a chance to realize capital gains and new investors a chance to share in future growth and profitability.

Maybe one of the most important conclusions of the examination of the progress achieved so far in the Romanian transition is the main goal of economic policies for 1994 and 1995, as stated in the Memorandum of Economic Policies of the Government of Romania: "... combining tight financial policies with accelerated privatization and enterprise reform, thus completing the stabilization phase of the reform process and laying the basis for sustainable growth."[34]

Notes

1. John S. Earle and Dana Sapatoru (1992), 'Privatization in a hyper-centralized economy: the case of Romania', *CERGE Working Paper*, no. 15, p. 4, Prague.

2. We did not take into consideration agricultural and non-agricultural co-operatives as both categories were in fact subject to the same central planning system as the state enterprises before 1990. The data represent a rough estimate because there were no such statistics on forms of ownership.

3. EBRD - *Private Sector Development in Romania*, Report to the First Meeting of the Consultative Group for Romania, Brussels, 10-11 May 1993, p. 2-3

4. When judging the results of the economic reform in Romania until now, we could agree with Laszlo Csaba about the consequences of the crisis-induced reform:

1   There is a significant discrepancy between reform theories and reality, because the latter is much more predetermined by legacies of the past and the inertia of institutions and structures than the theories that reflect a rejection of the actual state of affairs.

2   The reform is usually instituted usually by an already weakened government, which tries to avoid sharp clashes with the well-organized vested interests of the previously privileged groups and follows willy-nilly the popular mood.

3   Formal scenarios as approved by governmental committees and as translated into legislation lose much of their significance.

See Csaba, L. (1990), 'External implications of economic reforms in the European centrally planned economies', *Journal of Development Planning*, no. 20, United Nations, p. 108-111.

5.   Memorandum on Economic Policies of the Government of Romania (sent to IMF, but not published yet, as it was not yet approved by the Parliament), p. 1

6.   Because of the social and political implications of measures regarding the exchange rate, the National Bank of Romania:

> ... had to take into consideration the electioneering moment, with the intention of not affecting in any way the voters' option. From this point of view, the new attempt to freeze the exchange rate represented the price paid for the undisturbed continuation of the democratic process.

Rapport privind regimul valutar din Romania si masurile necesare pentru stabilizarea cursului leului (Report regarding the exchange rate regime from Romania and the measures necessary for stabilization of the exchange rate of the lei), National Bank of Romania, Bucharest, May 1993, p. 16

7.   For an English translation of relevant laws see Romanian Development Agency (1992), *Law Digest for Foreign Investors*, Bucharest, February.

8.   The rights of foreigners to own properties and to set up businesses were also separately ruled (Decree-law 96/1990 and then Law 35/1991) and, when the new Constitution was introduced, the provision that "... foreigners cannot own real estate in Romania ..." gave room for ambiguous interpretations (the foreign investment regulation gives national treatment to companies 100 per cent foreign owned, if registered in Romania, under Romanian law, but in many cases central and local authorities and even courts of law contested the right of such companies to own real estate).

9.   As an example, the Parliament approved in April 1992 Law 36 of the State Budget (with lei 1,031.5 billion as revenues and lei 1,120.5 billion as expenses), in July 1992, Law 62 regarding the Adjustment of the State Budget for 1992 (providing an increase of lei 196.5 billion both in revenues and expenses) and in February 1993, Law 11 for the Approval of the Government Emergency Ordinance 1/1992 regarding the Adjustment of the State Budget for 1992 (the budget recorded a lei 263.024 billion deficit compared with lei 88 billion deficit approved in the Budget Law).

10. 'Central and East Europe: Myths and Realities', *The Amex Bank Review*, vol. 20, no. 6, p. 6., 1993.
11. 'The state - an obstacle or a promoter of the market economy?', *Economistul*, May, 1992.
12. Article 8 of the Law on Salaries 14/1991 (Official Monitor no. 32 of 9 February 1991) provided that the Government, with the endorsement of the trade unions, may adopt measures against wage increases through extra taxes for a period of maximum one. See also Law 39/1992 (Official Monitor 73/23, Expres. 44, 2-8 November 1993.
13. A report to the Romanian Government done by Hay Management consultancy group.
14. For example, in Government Decision 388/1993 (Official Monitor no. 207 from 27 August 1993) regarding the approval of the merger between Avicola Covasna S.A. and Covasnacomb S.A. it is said that the two companies are totally state-owned. In accordance with privatization provisions, 70 per cent of their shares belong to the SOF and 30 per cent to the POF. For more confusion, the Government Decision regarding the organization of the Ministry of Industries states that the ministry is a shareholder in the subordinated companies.
15. See Bromley, D. W. (1989), *Economic interests and institutions*, Basil Blackwell Inc., New York.
16. See Drucker, P. (1992), *Managing for the future: the 1990s and beyond*, Truman Talley Books/Dutton, U.S.A.
17. In the public sector, under Law 15/1990 the process of 'commercialization' (corporatization) of the former state-owned economic units took place during 1990 and 1991, ending in April 1991 (all the state enterprises were transformed into corporations with the state as the only shareholder).
18. Hunya, G. (1991), 'Speed and level of privatization of big enterprises in Central and Eastern Europe: general concepts and Hungarian practice', *Working Paper*, no. 177, The Vienna Institute for Comparative Economic Studies, p. 4.
19. A very accurate description of the 'holding companies' and a well documented discussion of the related phenomena is given by Frydman, R., Rapaczynski, A., Earle, J. S. et al. (1993), *The privatization process in Central Europe*, Central European University Press, London, pp. 257-262.
20. The aide-memoire from the meeting of the Council for Coordination, Strategy and Economic Reform with private employers, which took place on 17 June 1993, contains such complaints and others, as follows:

    1   public administration employees do not respect the laws;
    2   Ministries' regulations are in conflict with the laws;
    3   state-controlled companies still receive quotas for some products (sugar, salt etc.);
    4   state-owned companies do not respect the contracts signed with private companies, do not sell their depreciated assets to the private companies and often are reluctant to trade with private companies.

21. Baltag, Vasil (6 January 1994), 'The industry sector and the economic reform', *Adevarul* (article signed by the secretary of state in the daily newspaper).

22. NAP, 1993.

23. Valencia, M. (1993-94), 'Too SOFt', *Business Central Europe*, December-January, p. 53.

24. Gabor Hunya stresses the fact that the fight against nomenklatura privatization slowed down the process both in Poland and Hungary, ibid., p. 9-10.

25. The Parliament is involved through the Privatization Committee of the Senate, the Committee for Industry, Trade and Services and the Commission for Economy, Budget and Finances of the lower Chamber.

26. The privatization regulation requires 'acts of transfer' signed by the Ministry of Finance, endorsed by the National Agency for Privatization, to be handed over to the SOF and to the POFs.

27. After the failure of the attempt to privatize the company within the 'Pilot' Privatization Program, starting late 1992, the management board of Tarom, the main national airline, with the support of the department for civilian air transportation, promoted a plan to issue supplementary shares, distributed to the employees and management, without a proper evaluation, thus trying to acquire a significant ownership stake in the company before its privatization; to be legally registered, such a move necessarily requires a written decision of the shareholders, but, on the grounds that the company is majority owned by the state, a court of law almost validated the action, substantiated only with the decision of the board of state representatives and the ministry's approval. The process was stopped by legal action taken by the Private Ownership Fund IV Muntenia, which holds 30 per cent of the shares in Tarom and there is still doubt whether this action is going to be successful as the document attesting to the transfer of shares was not in the possession of POF IV Muntenia when the increase of capital was initiated by the board of the company.

    Thirty per cent of the shares of Petromin, a large company that owns and operates about 300 vessels (most of Romania's merchant fleet) as well as port facilities, are held by POF III Transilvania (nota cu MO 77/1993) but the company's board, backed by the Minister of Transportation and the Prime-Minister, came out in July 1992 with a 100 per cent increase in capital, with the new shares paid in by a foreign (Greek) ship owner, not even bothering to notify POF III Muntenia of this action; this case, involving the control from the part of a foreigner of a company considered to be a significant national asset, fuelled a major debate that unfortunately concentrated on the issue of the right of foreigners to acquire majority interest in privatized companies and not on the many irregularities originated in the non-enforcement of the existing legal framework.

28. Earle, J. (1993), 'Why the SOF cannot function' and 'Privatizing the SOF: principles and explanations', *Discussion Papers for the Romanian Government*, July.

29. Government Decision of 26 February 1992 provides revaluation coefficients for 43 sectors of the economy. In effect, this decision - which is meant to result in values that reflect current replacement costs - has the effect of asking the investor

in an industry with high replacement costs to pay more for assets which he must then replace at further high cost because most have reached the end of their useful life or are technically obsolete. Conversely, the investor who can modernize his productive plant with a relative low investment receives a bonus in the form of a low acquisition price.

30. The memorandum on the policies of the Romanian Government attached to the last stand-by arrangement with the IMF provides for the privatization of only one bank.
31. A good review of the relevant literature concerning the neoclassical economics approach to property rights, economic organization, role of the state and the free-rider problem, in Eggertsson, Thrainn (1990), *Economic behavior and institutions*, Cambridge University Press, Cambridge.
32. ibidem, p. 12-13.
33. ibidem, p. 36-37.
34. See the memorandum as in note 5.

References

Balassa, B. (1989), 'Financial liberalization in developing countries', *Working Papers*, no. 55, The World Bank, Washington.

Coase, R. H. (1988), *The firm, the market and the law*, The University of Chicago Press, Chicago.

Commander, S. and Coricelli, F. (1990), 'The macroeconomics of price reform in socialist countries: a dynamic framework', *Working Paper*, no. 555, The World Bank, Washington.

Copeland, T. E., Weston, J. F. (1988), *Financial theory and corporate policy*, third ed., Addison-Wesley, Reading.

Country Economics Department (1991), *The transformation of economies in Central and Eastern Europe: issues, progress and prospects*, The World Bank, Washington.

Daiwa Securities Co. Ltd. (1991), *Proposal and issues related to implementation of the equity system*, Tokyo.

Fair, D. E. and Raymond, R. J. (ed.) (1993), 'The new Europe: evolving economic and financial systems in East and West', *Financial and Monetary Policy Studies*, no. 26, SUERF, Kluwer, Dordrecht.

Fischer, S. and Gelb, A. (1990), 'Issues in socialist economy reform', *Working Papers*, no. 565, The World Bank, Washington.

International Finance Corporation (1991), *Emerging stock markets factbook: 1991*, Washington D.C.

Jackson, M. (1991), 'Constraints on systemic transformation and their policy implications', *Oxford Review of Economic Policy*, vol. 7, no. 4.

Jackson, M. (1992), 'Practical, equity and efficient issues in the privatization of large-scale enterprises", *Working Paper*, no. 1, Leuven Institute for Central and East European Studies, Leuven.

Khan, M. S. and Clifton, E. V. (1992), 'Inter-enterprise arrears in transforming economies: the case of Romania', *IMF Papers on Policy Analysis and Assessment*, PPAA, IMF, Washington.

Lavoie, D. (1985), *National economic planning: what is left?*, Ballinger, Cambridge.

Masuoka, T. (1990), 'Asset and liability management in the developing countries: modern financial techniques - a primer', *Working Papers*, no. 454, The World Bank, Washington.

Milanovic, B. (1992), 'Distributional impact of cash and in-kind social transfers in Eastern Europe and Russia', *Working Papers*, no. 1054, The World Bank, Washington.

Milgrom, P. and Roberts, J. (1992), *Economics, organization and management*, Prentice-Hall, Englewood Cliffs.

National Agency for Privatization (1993), *The stage of the privatization process and proposals for its acceleration*, Bucharest, 23 December.

Paul, S. (1992), 'Institutional development in World Bank projects: a cross-sectoral review', *Working Papers*, no. 392, The World Bank, Washington.

Pearce, J. L. and Branyiczki, I. (1992), 'Revolutionizing bureaucracies: managing change in Hungarian state-owned enterprises', *Working Paper*, Graduate School of Management, University of California, Irvine.

Pearce, J. L., Branyiczki, I. and Bakacsi, G. (1992), *Person-based reward systems: a theory of organizational reward practices in reform - communist organizations*, paper presented at the Second Annual Western Academy of Management International Conference, Katholieke Universiteit Leuven.

Postolache, T. (co-ord.) (1991), *Economia Romaniei - secolul XX*, Ed. Academiei Romane, Bucharest.

Puffer, S.M. and McCarthy, D. J. (1991), *Integrating managerial decision-making authority within and between organizations: a Soviet-American comparison*, College of Business Administration, Northeastern University, Boston.

Rudnick, D. (1992), 'Short-term respite for Romanian banking', *Central European*, no. 16, October.

Schjelderup, G. (1990), 'Reforming state enterprises in socialist economies: guidelines for leasing them to entrepreneurs', *Working Papers*, no. 368, The World Bank, Washington.

Scott, D. H. (1992), 'Revising financial sector policy in transitional socialist economies: will universal banks prove viable ?', *Working Papers*, no. 1034, The World Bank, Washington.

United Nations (1992), *Report of the ad hoc working group on comparative experiences with privatization on its first session*, Geneva.

7 Delayed privatization and financial development in Bulgaria

Ognian Panov

Introduction

The focus of this chapter is the process of privatization and creation of capital markets in Bulgaria. Central to this study were the factors which influenced the specific method of transition from a centrally planned to a market regulated economy.

The capital market in the command economy was limited to a small private housing sector strictly controlled by the state. There were no financial instruments typical of the capital markets such as stocks, bonds, mortgages, etc. The money market was at a very early stage of development with the creation of commercial banks. There was no free foreign exchange and it appeared mostly as an irregular (black) market. In practical terms, the development of money and capital started from zero. Generally, this was a common phenomenon for most of the Central and East European countries with slight differences in the case of Hungary.

This chapter attempts to describe the main factors which influenced the specific method of transition in Bulgaria and caused the delay of privatization and specific development of the financial markets. To serve these purposes, it focuses on two main problems:

1   privatization and creation of a private business sector;

2   the development of new financial instruments and secondary markets and the role of money market development.

In order to understand the specific features of the Bulgarian transition to a market economy and the creation of financial markets, the political, economic and social environment is described. Internal political confrontation slows the process of reform and in the same way the process of privatization and market creation. The introduction of new financial markets passes different stages from spontaneous to regulated. In the paper illustrations are given of this phenomenon. The creation of a liberal money market in Bulgaria helps financial market development and this is the reason to look at its development.

Political, economic and social environment in Bulgaria

Democratization process

The Bulgarian way to democracy was a peaceful transition in the turbulent Balkan peninsula. Creation of democratic institutions was a result of two free general elections in June 1990 and October 1991, adoption of a new constitution in July 1991 and direct presidential elections in January 1992.

After the elections in 1991, when a five per cent barrier was introduced, a well-known western political model (Britain, Germany) with two major parties and small parties in between was set up in the parliament. The parliamentary parties were the UDF (Union of Democratic Forces, anti-communist coalition), the BSP (Bulgarian Socialist Party, the ex-communist party) and a small party of ethnic Turks, the MRF (Movement for Rights and Freedom). This model is still stable in the public opinion according to the monthly polls provided by Gallup International with roughly 50 per cent supporters of the big parties, 5-7 per cent for the Turkish party and the rest for the numerous small centrist parties with less than 5 per cent for each and a growing number of people not voting. This model could be very effective if there was not the existing controversy due to the ex-communist party remaining unified and very powerful and new democratic forces with no ability to keep the unity. Thus in the parliament a new fourth group of former members of the UDF appeared who were separated mainly on the basis of personal misunderstanding. In this model the role of the small party MRF is crucial. Political support for the governments changed practically on an annual basis. The first two were formed by Loukanov (BSP) in 1990. The third was headed by Popov (independent) in 1991 with partial support of all present in the parliament parties. This was very important for the start of economic reform in 1991. In 1992 the government was formed by Dimitrov (UDF) after the elections in October 1991. In 1993 Berov (independent) made a new government under the mandate of the MRF and with the support of the fourth parliamentary group separated from the UDF called Union for New Democracy (UND) and the BSP. This is also a coalition government which came to power with the mandate to start mass privatization. Because there are no pre-term elections in the spring of 1994, this government was the first with a two year mandate and more possibilities to move economic reform faster, but, unfortunately, so far without success.

The most publicly accepted institution is the presidency. President Jelio Jelev, the well-known dissident and the first leader of the UDF, plays the role of father of the nation and always has the support of the majority in public opinion polls and is constantly in first place for personal popularity. The presidential institution, according to the constitution, is not an executive body and does not have much power but plays at present a stabilizing role in society.

The activity of the legislative body, the People's Assembly, is far below what was expected and needed. By the spring of 1993, very important economic regulation laws were still not passed, such as the law for bankruptcy and the law for mass privatization, which were also the main conditions for further support of the IMF and the agreement on Bulgarian debts with the creditors from the London (private) and Paris (government) clubs. Many of the laws which establish monopolism and are against the new constitution still remain unchanged, such as the law for insurance. There are no new administrative and penalty codes which are very important for the creation of a new environment based on democratic principals. Thus, because of the controversial equilibrium in the political arena, reform is slowing down which causes the delay of major structural changes in property rights and economic behavior of firms. Political stalemate situations, like in chess, prevent effective measures against the fall of the national economy and support the creation of a large-scale shadow economy mixed with the regular one. That is why the budget deficit is more than predicted by 5-6 points and collection of taxes less than 70 per cent. The biggest success of the legislative body and the government is the adoption of the VAT law with the single rate of taxation as was recommended by the experts of the IMF and the EU. From 1 April 1994, taxation is effective with an 18 per cent rate and a zero rate for some important basic goods such as bread, energy and baby food. This creates a basis for better control on chains of trade and makes the economy transparent. But the conflict with the editors of newspapers who consider this law as a killer of the free press shows that society is still not prepared for deep changes in the economy. When the process of democratization is practically fulfilled, the process of economic reform is slowed down.

In the present political situation all political forces declare that they are in favor of deep economic reform. But none of the governments did much to move ahead the most important part of the structural economic reform process, privatization. After the first free elections in 1990 and later in 1991, in Bulgaria all possible combinations for the political support of the government were tried: BSP alone, UDF alone, unofficial coalition of both parties, until the MRF imposed a government with the support of the BSP and part of the UDF. That shows the lack of interest and political will in all the political forces and in society the desire for rapid privatization. The reason for that we shall try to discover later in this chapter.

Development of the capital market in Bulgaria took place not only in a passive political environment but, what is a more important obstacle, in a deep economic crisis. Statistical data from different sources or from the same sources but in different periods differ greatly, but the picture is in general the same: fall in production, fall in the gross domestic product (GDP), high inflation and growing unemployment. In 1993 this process was still slowing down but there were no firm indications for a turn to growth. For illustration we use data from the annual report of the BNB for 1992 and the already approved but not yet published report for 1993 given in tables 7.1, 7.2, 7.3 and 7.4.

The unprecedented fall in production was equally heavy in all sectors of the national economy. The gross output fell 15-20 per cent annually. The most affected branches were those which had been oriented mostly to the former Soviet market. Manufacture of machinery, for example, dropped at the end of 1992 to 70.4 per cent below the level of the last quarter of 1989. Because of the delay of privatization during this entire period more than 90 per cent of industrial assets remained state-owned.

The industrial sector was hit not only by the decline in demand, but was also affected by financial burdens. Most state-owned enterprises are in a financial fix, many of them heavily burdened with long-term credits inherited from the communist period which they are unable to service at the high nominal interest rates. The corporate industrial sector, state-owned and cooperative, closed 1992 with an aggregate loss of leva 7.1 billion (3.6 per cent of official GDP). In the first quarter of 1993 alone, this aggregate loss was leva 9.4 billion with practically all industrial branches, with the exception of the tiny printing industry, operating at a loss.

Output of agriculture also declined in the period under consideration, mainly due to institutional rigidities associated with the process of agrarian reform, the maintenance of price controls on agricultural products, and the lack of finance. Agrarian reform was started in 1991 by the adoption of a bill of restitution of agricultural land to previous owners (attended in 1992). This act presumes the liquidation of the existing Soviet-type agricultural cooperatives and disposition of their assets (including land, livestock, machinery, etc.) to previous owners or their heirs. However, the actual process is cumbersome and lengthy, besides, the new owners are inexperienced and undercapitalized. Bad loans and uncertainty about the duration of their existence have deprived cooperatives of access to commercial credits. For 1993 state-owned banks were obliged by law to grant loans financing current agricultural operations, involving indirect subsidies to agriculture. Livestock breeding was most seriously affected: in the period 1991-92, the number of cattle and pigs decreased by over 25 per cent and sheep by over 35 per cent. Crop production fluctuated depending on weather conditions: a relatively good grain harvest in 1991 was followed by a sluggish one in 1992.

In this situation the constant fall of GDP in private consumption and retail sales and in gross fixed investment is understandable, which describes a full picture of total economic crisis. Development of a new capital market for the country is a very slow process because of the dwindling commodity market. External input which might help the development of the national economy was very limited.

Table 7.1
Basic economic indicators for the years of transition
(mln. leva - current prices)

Indicators	1989	1990	1991	1992	1993*
GDP produced	39,579	45,390	131,058	195,000	286,134
Industry and construction	23,507	23,273	62,843	90,800	119,258
Agriculture	4,331	8,055	20,139	20,200	26,273
Services	11,742	14,062	48,076	84,000	135,170
GDP used	41,103	47,385	135,916	209,480	311,467
Final consump.	27,999	33,585	99,305	164,170	266,240
Population	25,180	30,298	78,759	130,640	203,986
Government expenditure	2,918	3,286	20,546	33,530	62,254
Gross accumulation	13,105	13,800	36,611	45,310	45,227
Capital investment	10,328	9,652	24,777	39,310	33,363
Changes in inventory	2,776	4,148	11,834	6,000	11,864
Foreign trade balance, losses and stat. discrepancy	-1,524	-1,995	-4,858	-14,480	-25,333

* preliminary data
Source: BNP, *Annual Reports,* 1992, 1993.

Table 7.2
Basic economic indicators for the years of transition
(rate of change - %)

Indicators	1990	1991	1992	1993*
GDP produced	90.9	88.3	94.3	95.8
Industry and				
construction	87.5	81.4	89.0	96.1
Agriculture	96.3	107.7	92.3	83.7
Services	95.7	88.7	96.7	94.0
GDP used	90.9	88.3	92.3	-
Final consumption	100.6	85.6	92.3	97.2
Population	98.5	84.3	92.6	100.3
Government				
expenditure	108.6	89.8	90.9	87.8
Gross				
accumulation	74.9	84.4	72.1	78.4
Capital				
investment	81.5	80.1	98.5	-
Changes in				
inventory	50.3	94.3	16.9	-

* preliminary data
Source: BNP, *Annual Reports,* 1992, 1993.

Table 7.3
Production by economic branch at 1990 prices
(mln. leva)

Branch	1989	1990	1991	1992
Industry	59,625	52,747	39,957	31,709
Construction	6,038	6,158	5,983	4,748
Agriculture	14,852	13,815	14,434	13,560
Transport	5,067	4,578	4,481	3,885
Trade	4,155	4,445	5,370	3,666
Communications	757	840	910	1,230
Other	1,258	1,198	1,388	442

Source: NSI.

Table 7.4
Production by economic branch at 1990 prices
(index - 1989 = 100 %)

Branch	1990	1991	1992
Industry	88.5	67.0	53.2
Construction	102.0	99.1	78.6
Agriculture	93.0	97.2	91.3
Transport	90.3	88.4	76.7
Trade	107.0	129.2	88.2
Communications	111.0	120.2	162.5
Other	95.2	110.3	35.1

Source: NSI.

Foreign trade and investments

Reorientation of foreign trade after the failure of Comecon is slowly taking place with the prevalence of imports over exports. Within only two years, the share of exports to Central and East European countries has dropped twice from 80.2 per cent in 1990 to 41.9 per cent in 1992 (table 7.5).

Table 7.5
Export and import by groups of countries (relative share in %)

	1985	*1990*	*1991*	*1992*	*1993*
Export	100.0	100.0	100.0	100.0	100.0
from which:					
OECD countries	9.6	9.0	26.3	42.3	43.1
including:					
EEC countries	6.4	5.0	15.7	30.8	28.1
Asian countries	1.6	1.5	3.4	3.3	3.0
Central and East European countries[*]	74.7	80.2	57.7	41.9	36.4
Arabian countries	9.6	6.1	8.3	7.9	6.9
Import	100.0	100.0	100.0	100.0	100.0
from which:					
OECD countries	15.3	15.0	32.8	46.5	42.6
including:					
EEC countries	9.4	9.6	20.7	32.6	30.2
Asian countries	3.6	3.2	7.8	6.7	5.7
Central and Eastern European countries[*]	74.9	75.9	48.5	37.2	45.2
Arabian countries	4.6	4.3	4.5	7.7	4.9

[*]   Albania, Poland, Romania, Hungary, ex-USSR, Czechoslovakia and Yugoslavia. Ex-DDR is included for 1985 and 1990.
Source: NSI, BNB.

The share of OECD countries has risen significantly from 9 per cent in 1990 to 42.3 per cent in 1992 but in absolute value terms in USD the increase was very small. In 1992 imports from OECD countries were greater than exports by $ 130 million which put additional pressure on national industry. The delay of the trade agreement with the European Community in 1993 will make the situation even worse.

The main financial problem of the Bulgarian economy is foreign debt. Gross debt rose from $ 9.2 billion at the end of 1989 to $ 12.9 billion at the end of 1992. In March 1990 Bulgaria announced a moratorium on debt services which was a blow to foreign trade relations. Only recently the agreement with the London Club (private creditors), at a preliminary stage at the end of 1993, gives some prospects to improve the situation with the Bready-type agreement which reduces the obligation to the London Club banks by close to 50 per cent. This is a great success of Bulgarian economists which may significantly change the foreign investment situation and will create new opportunities for capital market development.

Foreign direct investments in the Bulgarian economy since 1989 are disappointingly small. Their cumulative level for the period 1990-92 is 34 times less than in Hungary, which is a country with relatively the same size and population. Only 836 joint ventures were registered in the Bulgarian Chamber of Commerce as of April 1993.

Some recent reports cast more light on the problems with foreign investment in Bulgaria. In February-March 1993 the Center for the Study of Democracy in Sofia conducted a poll among the largest foreign investors currently present in Bulgaria.[1] The results of this poll show the four most important factors which negatively affect the inflow of FDI:

1   the unresolved foreign debt problem;

2   slow privatization;

3   the underdeveloped banking system;

4   unstable legislation.

As regards the future, 64 per cent of the respondents defined their business plans in Bulgaria as long-term, 29 per cent as medium-term and 7 per cent as short-term.

The Deloitte Touche consulting company also conducted a recent survey of investment conditions in Bulgaria.[2] According to its findings the most serious problems faced by foreign investors (apart from the problems associated with foreign debt) are:

1   poor infrastructure, especially concerning telecommunications;

2   suppliers' problems (low elasticity of supply to changing market conditions);

3   finding data on market potential as well as legal information (lack of reliable, public information services);

4   an inadequate banking system;

5   regulatory issues (e.g. the fiscal system is in a transitional state);

6   inflation and high interest rates.

The study provides several breakdowns of foreign investments:

1   by age: 25 per cent - less than 1 year; 42 per cent - from 1 to 3 years; 33 per cent - over 3 years old;

2   by size: 67 per cent - less than $ 1 million; 25 per cent - from $ 1 to 5 million; 8 per cent - over $ 5 million;

3   by sector: 50 per cent in trading; 25 per cent in other services; 25 per cent in manufacturing.

Social framework

Social development in Bulgaria is heavily influenced by the high rates of inflation and unemployment (table 7.6). The shock therapy in Bulgaria was related to the liberalization of prices in February 1991. The annual rate of change of the CPI by the end of the year was 473.7 per cent. This shock was tempered by the relatively large reserves of food common for Bulgarian households. At the end of 1992 the inflation rate was 79.5 per cent and in 1993, 62.5 per cent. This tendency of reducing inflation is one of the reasons for the relatively calm social situation with strikes only in certain branches of industry like mining, machine-building, etc., severely affected by the loss of jobs. The new budget for 1994 is based on the forecast of a 40-45 per cent inflation rate based on the CPI but sharp devaluation of the leva at the beginning of 1994 and introduction of VAT may double this figure.

The gap between CPI and PPI in recent years might create problems (Miller, 1993). It is a result of the excessively long mediators' queue between state enterprises and consumers. Only the privatization process may change the situation.

Unemployment was the other soft spot of the transition process. It rose very rapidly from practically zero in 1988 to 18.6 per cent of the labor force at the end of 1992. In 1993 the process was slowed down and in September-October the first reduction of the unemployment level occurred with a two per cent drop.

Table 7.6
Labor, incomes and prices

	1990	1991	1992	1993
Consumer price index (growth rate in %)	50.6	473.7	79.5	63.9
Retail price index (growth rate in %)	64.3	317.4	78.9	-
Producer price index (growth rate in %)	330.5	40.2	26.9	
Average monthly wages and salaries (SOEs and coop.; nominal in leva)	361	963	2,102	3,000
Total incomes of private households (average monthly per cap.; nominal in leva)	259	693	1,337	1,985
Total employment (yearly average in thousands; public sector)	3,824	3,205	2,653	2,294
Empl. index 1989 = 100 %)	87.6	73.4	60.8	52.5
Unemployment (total) (endperiod values in thousands)	65	419	577	626
Rate of unemployment (%)	1.6	11.2	15.2	16.4

Source: Statistical Yearbook of Bulgaria; *Statisticheski Izvestia of the National Institute* (Quarterly Statistical Report); *Iznos i Vnos* (Exports and Imports, Quarterly bulletin of the NSI); News Bulletin of the Bulgarian National Bank.

For state employees, people on pensions and social security allowances, most painful was the rising budget deficit which was 8.4 per cent in 1990, 4 per cent in 1991 and 7.1 per cent in 1992 on cash basis from the GDP. At the end of 1993, the government had to break its previous decision coordinated with the IMF for a six per cent barrier of the deficit and the probable rate of the budget deficit will be close to seven per cent of GDP.

The main reason for the deficit appearance is the failure of the government to collect taxes. In November 1993, only 67 per cent of the budget input was collected. Main debtors are the state-owned enterprises. Thus, the privatization process might help to solve problems of groups of people who count on budget subsidies.

Creation of the private sector in Bulgaria as a source for primary capital markets

The transition to the market economy cannot be realized without the creation of market agents. That means households should receive property rights on all kinds of resources (in addition to labor), and new private firms should be created, independent from the state bureaucracy. In addition, restructuring and commercialization of state-owned enterprises gives equal opportunities for market behavior of the state sector of the national economy.

Privatization process

The only method of development in all economies in transition is a large privatization process in combination with creation of new private companies. In the case of Bulgaria, the privatization process was stopped at the very beginning in August 1990 with the special decree of the Great National Assembly and has not yet started on a large scale. On the other hand, the development of new private firms was very rapid. But that created an imbalance which may be dangerous for future economic development.

The privatization story in Bulgaria has until now three major parts. The first which we can entitle 'spontaneous privatization' started right after 10 November 1989 and ended in August 1990 with the moratorium on privatization decisions. During that period there was no special privatization law. That situation was used by some heads of state enterprises, officers in ministries and local authorities to start privatization in their own interest on the basis of existing rights of state enterprises to sell parts such as workshops, stores, etc. or rent them for long periods with the permission of the ministries. Without real regulation of this process, the advantage of ruling groups at that time was obvious. This is the reason for the moratorium accepted by all political forces. The second period ends with the adoption of the Law on Reincorporation and Privatization of State and Municipal Enterprises on 23 April 1992 and can be named 'hidden privatization'. Mostly banks were affected by these privatization efforts because the banking law was adopted only a month earlier, in March 1992. Four banks which used to be branches of the BNB or state-owned trade banks were practically privatized with illegal raising of the stock capital of the banks. The third period which has still not ended could be described as a 'slow start'. Basic information on privatization activities is given in table 7.7.

The most striking result is the great difference between the decisions for privatization contracts. Only 14 per cent of the decisions were realized during the observed period. This is not a result of the long process of realization or late decisions but mostly of the change in mind of the decision-making bodies. One parallel with the previous study at the end of July 1993 shows that the decisions taken before 30 July 1993 numbered 589 compared to 457 before 31 October 1993 or only 30 per cent more. These are mostly local authorities' decisions and the change was motivated with preferences for short-term rent and then for privatization. This is one of the indicators for the lack of real public support for the privatization process. Practically all realized objects are in the trade sector (60 per cent) and industry (23 per cent). This shows the limited sphere of interest for privatization.

Desperately low levels of privatization, only 3.5 contracts per month, and the small scale of privatized objects grounded the conclusion that privatization in Bulgaria has not started yet.

Financial data for the privatized objects is given in table 7.8. It shows that most of them are medium-scale: leva 33 million on average for state-owned and only leva 1.2 million per object for small-scale (at the end of October 1993, $ 1 was leva 26). The low level of payments for signed contracts, only 22.3 per cent, was a result of the privatization techniques. Half of the privatized firms are on the basis of debt-equity exchange or rent with obligation to buy after a certain period of time.

The costs of the privatization contracts are moderate: 12.8 per cent of the price of the object. The major privatization technique used was auction, for 67.2 per cent of the cases. Only 4.7 per cent of the objects were privatized with direct negotiation. Only three of all objects were bought by foreign investors. New investment is included in the contracts as a condition only for 12.5 per cent of the privatized objects.

Change in this situation was expected after June-July 1994 when the law for mass privatization was expected to pass the National Assembly. This would create greater public interest but the financial role of mass privatization will remain relatively limited.

197

Table 7.7
Privatization on legal basis (1 May 1992 - 31 October 1993)

Sectors of national economy	Approved for privatization	Privatized objects	Approved for privatization	Privatized objects
	Number		Structure %	
	1	2	1	2	1	2	1	2

Sectors of national economy	1	2	1	2	1	2	1	2
Total	175	282	6	58	100.0	100.0	100.0	100.0
Industry	79	65	5	10	45.1	23.0	83.0	17.3
Construction	13	1	-	-	7.4	0.4	-	-
Agriculture	4	13	-	1	2.3	4.6	-	1.7
Transport	51	13	-	-	29.1	4.6	-	-
Trade, material-techn. supply and buy up	19	148	1	38	11.0	52.5	16.7	65.5
Other sectors of material prod.	2	6	-	6	1.1	2.1	-	10.3
Housing and public utilit.	-	6	-	2	-	2.1	-	3.5
Science and science service	4	1	-	-	2.3	0.4	-	-
Health, social ins., sports and tourism	2	2	-	-	1.1	0.7	-	-
Other	1	27	-	1	0.6	9.6	-	1.7

1   trade companies
2   non-reformed enterprises and autonomous parts
Source: NSI.

Table 7.8
Total number of newly founded enterprises on 1 September 1991 and 24 February 1992 in sectors

Sector	1 Sep. 1991	Share in total number-%	24 Feb. 1992	Share in total number-%
Total	130,102	100.0	186,788	100.0
Industry	9,120	7.0	14,532	7.8
Construction	5,061	3.9	6,768	3.6
Agriculture	7,664	5.9	13,609	7.3
Forestry	71	0.0	88	0.0
Transport	20,879	16.1	26,633	14.3
Communication	25	0.0	38	0.0
Trade	32,311	24.8	47,451	25.4
Other sectors of material production	28,984	22.3	41,691	22.3
Housing	8,541	6.6	13,452	7.2
Science and service	1,235	1.0	1,427	0.8
Education	404	0.3	563	0.3
Culture and arts Health, sports	1,546	1.2	1,893	1.0
Finance, credit, insurance	141	0.1	181	0.1
Administration	89	0.1	123	0.1
Other sectors of non-material production	13,054	10.0	16,902	9.0

Source: *168 Hours,* no. 27, 1992.

The creation of the Bulgarian private sector started in 1989 when Decree 56 was adopted. Under this legislation, for three years (1989, 1990, 1991) more than 200,000 firms were created. This figure is much higher (three times) than the number of the firms in Hungary, a country that is similar by size and population to Bulgaria and which started its economic reform earlier. This shows the entrepreneurial approach of Bulgarian people which is essential for the transition to a market economy.

The state sector during that period was developing new business structures as a result of the process of demonopolization which began in 1991. From 913 state-owned firms at the end of 1989 there were 5,523 at the end of 1992, excluding the agricultural sector.

The overconcentration of business in the agricultural sector in 1989 at 332 agro-industrial complexes, turned to a completely new business structure which is still unclear because the process of restitution is not yet completed. More than one million private farms and several thousand cooperatives are expected. In 1990 the agro-industrial complexes were dissolved and were formed into more than 2,500 cooperatives. In 1992 the cooperatives were announced in liquidation under the new Cooperative Act adopted in 1991. But by the end of 1993 only about 30 per cent of the land had been given to the owners with legal acts.

Restitution is a start for the formation of a land market. In the Land Act approved in 1991, there was a limitation of the free sale of land for a three year period after the restoration of the rights on property. Immediate sale of land is permitted in the new Land Act of 1992.

The formation of joint ventures was the most appropriate method of penetration of foreign capital within the former socialist countries before the transition had started. The history of foreign joint ventures in Bulgaria began in the early 1980s when under Decree 545 the first JV was formed with a Japanese firm. At the end of 1989, the number of JVs in Bulgaria was 44 and at the end of 1991 it was already 850.

Under the new Trade Law adopted in 1991, new firms with foreign capital have the same organizational form as Bulgarian ones. There is no special calculation of JVs but the number of firms with foreign participation at the end of 1992 according to the NSI is 1,500.

A study conducted in Bulgaria in the first half of 1992 by the Institute for Small and Medium-size Enterprises of the Netherlands[3] gives a general idea on the private firms. Two-thirds of all firms registered in Bulgaria by 24 February 1992 are in trade, transportation and other areas of industrial services (table 7.8). Only 7.8 per cent of the firms have industrial purposes. This tendency of creating mostly non-industrial firms became stronger in 1991 and 1992 (table 7.9). The distribution of the private firms is relatively equal to the territory of the provinces with a peak in the Plovdiv region (table 7.10).

Table 7.9
Year of foundation and main field of activity

Start in	Production % firms	Material service	Trade	Other service	Total
Total	100	100	100	100	100
1989 or 1990	58	46	32	39	42
1991 or 1992	42	54	68	61	58

Source: *168 Hours,* no. 27, 1992.

Table 7.10
Number of private enterprises to 1,000 citizens in the region

Region	Number
Bulgaria	20.7
Sofia-city	20.8
Bourgas	21.4
Varna	21.0
Lovetch	21.8
Michailovgrad	14.4
Plovdiv	27.6
Razgrad	15.2
Sofia-area	20.7
Haskovo	18.5

Source: *168 Hours,* no. 27, 1992.

A very important issue is the analysis of the source of capital for the formation of the firms. The cited survey shows that more than 80 per cent of the capital comes from personal or family savings (table 7.11). It is important to say that 70 per cent of the firms are individually-owned and the average size of one firm by number of employees is 3.4 and by sales only leva 35.1 thousand or $ 1,500 (table 7.12). The very small size of the private firms makes them non-competitive on the capital market.

Table 7.11
Main source of foundation capital for firms (by towns)

			Town			
Source of capital-	Sofia	Varna Plovdiv	Medium towns	Small towns	Villages	Total
Total	100	100	100	100	100	100
Own savings	60	58	64	59	54	60
Family/friends	21	21	19	27	26	23
Credit from:						
another firm	8	12	14	13	18	13
bank	9	7	-	-	1	3
inapplicable	3	2	3	1	-	2

Source: *168 Hours,* no. 27, 1992.

In 1993 for the first time the National Statistical Institute published estimates of the share of the private sector in the Bulgarian economy. According to these figures in 1992 the private sector accounted for about 16 per cent of the country's GDP (43 per cent in agriculture, 13 per cent in manufacturing and 12 per cent in services); 52 per cent of trade (domestic and foreign) is already in the private sector.

Table 7.12
Private enterprises in sectors for 1990

Sector	Number of firms	Number of empl.	Average size	Incomes (mln. leva)	Expenses (%) to incomes
Material production	40,382	124,076	3.1	1,509	34.6
Industry	14,331	42,657	3	383	79.7
Construction	3,022	28,034	9.3	283	86.4
Transport	10,379	15,536	1.5	75	75.7
Trade	8,457	25,407	3	640	88.1
Non-material production	7,592	39,302	5.2	174	56.6
Services	4,393	30,093	6.9	82	27.9
Education	306	906	3	6	30.7
Culture	745	2,477	3.3	17	88.3
Health, sports, tourism	378	1,215	3.2	27	39.7
Total	47,974	163,378	3.4	1,683	81.7

Source: NSI.

According to estimates of the Agency for Economic Coordination and Development, the level of employment in the private sector in 1992 was about 470,000 persons. This would make up for a total employment of 3,030,000 persons, or 270,000 over the officially recorded number. Total labor force would also be higher by the same margin. This would reduce the rate of unemployment to just over 17 per cent, still above the official figures.

The development of the private sector in Bulgaria is not an alternative to rapid and mass privatization. The private sector for the last four years has only grown ten points in GDP (table 7.13).

Table 7.13
Share of the private sector in GDP

Indicators	1990	1991	1992	1993[*]
Private sector-total	9.1	11.8	15.3	19.4
Agriculture and forestry	6.0	5.5	5.7	5.3
Industry	1.9	2.8	4.0	4.9
Services	1.2	3.5	5.6	9.2

* preliminary data, computed on current prices
Source: NSI, BNB.

This growth is based mostly on the development of the private service sector. The delay with restitution and land reform did not permit the development of the private sector in agriculture. Thus, the strategic political decision to give priority to restitution versus privatization practically stopped the structural changes in ownership. With less than 20 per cent of the GDP, the private sector does not strongly influence the creation of a competitive market. Only mass privatization can change the situation.

The development of the privatization process in Bulgaria is full of paradoxes. The first one is political support. All political parties present in the parliament support the privatization process. The latest government of Luben Berov was announced as a government of privatization. Nevertheless, until the end of 1993 there was only one successful operation of the so-called big privatization: the Belgian company 'Amylum' bought the new non-operating plant 'Zarevichny producty' in Razgrad.[4] In a TV interview the head of the Bulgarian Privatization Agency, Renata Indjova, declared that there is an institutional opposition to the privatization process, but did not specify any institution.

The second paradox is in the political parties' programs. They stated that privatization was a first priority because of the very large state sector at the beginning of the transition period containing more than 90 per cent of fixed assets. But priority was given to restitution. In February 1991 the Land Reform Law was passed in the BSP-dominated parliament and resumed the restitution of land to previous cooperation land owners.

The next restitution law was passed in the period December 1991-January 1992 and regulated the restitution of urban real estate and industrial properties nationalized in 1948 to previous owners or their heirs.

Finally the privatization law was passed in April 1992. The last two laws were adopted by the parliament under the domination of the UDF. The priority for restitution created many small owners and a new class of rentiers but did not help the structural change in the national economy.

The third paradox is the very strong social orientation and populistic appeals both of the BSP and the UDF but none of these powers accepted in the beginning the mass privatization model which was developed in Czechoslovakia and was implemented also in Poland and Russia. In October 1993 there was a complete proposal for the implementation of a system similar to the voucher system of bonds with delayed payment but a decision is not likely to be taken soon. According to the present law the main privatization techniques are auction, open competition, negotiation with potential buyers and sale of stocks. Personnel of an enterprise have the right to buy not more than 20 per cent of enterprise stock in direct payment. Thus obtained stock does not give voting rights.

The next paradox is the wish to make the process of privatization transparent with clear structure and division of rights between the institutions responsible for privatization and the announcement of auctions, official competitions and, sometimes, numerous scandals about hidden privatization which makes the public suspicious of it. Many of the auctions for small and medium-size enterprises failed because of a lack of buyers. Usually the first round fails because there is a rule to reduce the price on the next rounds up to 30 per cent. But the main reason is that buyers prefer direct negotiations. The public relates this kind of privatization to corruption.

This situation creates a specific stress in the institutions which may be called 'privatization complex'. The realization of privatization contracts is very prestigious but on the other hand it always creates problems which might turn into public scandals terminating with the replacement of responsible people. That is why the most appropriate formula for them is to make programs for privatization and to delay as much as possible their realization.

'Privatization complex' has a background in the first steps of irregular privatization which started early in 1990 on the basis of existing rules of Decree 56. A part of the old nomenclature tried to privatize and convert activities of the existing state-owned enterprises to the newly created private firms of their own or of some relatives. This was common for touristic objects and some retail trade objects using long-term lease contracts with unfavorable conditions for the state. During this period a major scandal occurred, which is not cleared yet, with the hidden privatization of some hotels in one of the most popular sea resorts 'Sunny Beach'. The local authorities, as well as government bodies were involved. This scandal was the immediate reason for the moratorium on privatization in August 1990.

Nevertheless until now there have been many accusations of hidden privatization. It was found in the commercial bank sector. Large banks like the Bank for Agricultural Credit, First Eastern International Bank, and Bank 'Biochim' were mentioned in this context. Critics caused special investigations and the white paper by the BNB on the issue in the fall of 1993. There was no direct result but it is clear that some of the banks became private without a special decision for privatization of any governmental institution.

Failure of the recant program for large privatization in 1993 provoked additional suspicion about the method of preparation for privatization. The most recent case is the privatization of six cement plants which were decomposed into six separate firms in place of the old SOE. In November 1993 it was announced that the contract between the government and the consulting firm 'Banker's Trust' was imperfect, too expensive and the consultant did not fulfill his obligations in favor of the government. Thus the largest privatization program in size was stopped with no prospect for the near future.

Because of several changes in the management body of the agency for privatization there is no accepted new program for privatization for 1994. This is a result of internal political fights which will influence the decision-making process. As a result of this the privatization has not started yet.

The most important obstacle for privatization, in our understanding, is the lack of interest in the management of the state-owned firms. In the existing law there is no special procedure for management buy-out (MBO) and the financial possibilities of the managers are not very big until now. But, what is more important, the delay of privatization is a source for the creation of substantial personal wealth for them without big risks.

Miller (1993), who has studied this problem, has a very serious explanation for this phenomenon:

> State enterprise managers do not share directly in the profits of the enterprise; therefore, they do not have the same motivation to increase profits as managers in private firms. Furthermore, the new environment presents many challenges for these managers. For many firms demand for their products has fallen sharply as exports to the former Soviet Republics have declined. Financially they are in serious difficulty and are dependent on loans from the banks for their survival.

The managers also face an uncertain future. The Law for Privatization which has passed in the spring of 1992 sets out a procedure for privatization of large enterprises, but actual privatization has been very limited. Managers, therefore, face a diverse set of possibilities: continuing to manage a state enterprise for some time to come, managing a privatized firm, an ownership role in the private enterprise, or having no further role in their present enterprise.

Evidence as to what managers are actually doing is scarce and anecdotal. Given such large future uncertainties, however, it would not be surprising if managers were short-sighted and attempted to exploit their present position at the expense of the future well-being of the enterprise. For example, decapitalization of the firm may create present profits, but lower its future value.

Another concern is that managers may be establishing 'shadow firms'. These shadow firms are private firms where management of the state enterprise has a stake. These shadow firms might be operated by friends or relatives of the management. The state enterprise sells its output to the shadow firm, and the shadow firm then sells the product to others, effectively privatizing the profits of the state enterprises.

This explanation is indirectly proved by the phenomenon of the gap between CPI and PPI indexes which is very big and growing in Bulgaria (table 7.14).

Table 7.14
Retail/consumer and producer price indexes in Bulgaria
(January 1991 = 100)

	Jun. 91	Sep. 91	Dec. 91	Mar. 92	Jun. 92	Sep. 92	Dec. 92	Mar. 93	Jun. 93
R/CPI	376	451	515	597	723	793	912	1,070	1,219
PPI	325	361	394	416	454	478	492	524	551

Source: National Statistical Institute and Agency for Economic Coordination and Development. Through 1992 the retail/consumer index is a retail goods index. In 1993 it is a consumer price index.

There are no other reasonable explanations for such a large gap which in June 1993 is more than the value of current PPI. Management is interested in reducing the production price and shadow private firms are interested in receiving the highest possible price. In the situation of relative deficit and high prices of import products the real gap of individual products may be a substantial source for profit. This profit is distributed illegally between institutions and politicians creating an environment not in favor of real privatization. This situation might change with the concentration of capital in the hands of those participating in the distribution of shadow profit groups and further decapitalization of state enterprises. That may occur not earlier than the end of 1994.

Financial markets development

In the transition period the first free markets to appear were retail and business service markets, followed by wholesale markets. There was previous experience in retail markets with the so-called cooperative markets where production from agricultural land given for individual needs was mainly offered. The boom of street markets in the first years of freedom created a need for wholesale produce to sell by mostly unregulated individual imports and distribution among thousands of individual retail sellers. The process of restitution and small privatization helped to establish the modern infrastructure of retail sales which rapidly changed the external appearance of big cities.

Financial markets development met many more difficulties but the creation of the banking infrastructure also influenced the new image of big and even small cities. The main obstacles to the creation of viable financial markets were:

1   the lack of experience. The exchange system in the CMEA was based mostly on barter principles and physical plans. Financial planning was designed to play a subservient role to physical planning;

2   investment was not dependent on household savings but on centralized funds. With the creation of commercial banks state financing was changed to long-term credits which created the problem of bad debts still unsolved in the Bulgarian context;

3   very large external debt which was not served from early 1990. The moratorium limited the inflow of foreign investors. None of the foreign banks had a branch office in Bulgaria as of early 1994;

4   the delay of privatization limited the capital market only to a small number of private stock companies and private banks.

The creation of financial markets in Bulgaria passed different stages of development which are inevitable in the political and economic situation of transition. The first one is the stage of spontaneous markets. The most aggressive people who have the opportunity to break the existing totalitarian laws without fear of being punished start illegal but open operations in some fields of money and capital markets, mostly foreign currency exchange. At the very beginning, before the political failure of the totalitarian regime, dealers were punished in some cases. But the premonition of the new era gave them courage and made the official institutions careless. The boom of the spontaneous markets came in the first years after the political change. It disappeared slowly under the pressure of the new market institutions.

Spontaneous markets appear when there is a great need for exchange but state regulation delays it. Market agents exist but there are no rules. Many of the old rules, for example, for currency exchange, were not replaced with new legal regulations but simply were not applied in practice.

Another example is the insurance market. The old legislation recognized only the monopolistic insurance agency State Insurance Institute (SII). But for the last four years three other state-owned insurance companies and six private insurance companies with foreign capital were divided from it. Foreign, mostly Austrian, insurance companies offer their products on the life insurance market. There is no legal regulation on insurance activities even on the executive level. The law on insurance is still waiting in the parliament but the market operates without regulation.

The second stage is the introduction of instruments and institutions by the state using the example of developed market economies. In this situation, government institutions and the state-owned firms and banks play the role properly in some sense but they have neither the knowledge nor the interest to be real market agents. Lack of legislative acts and control over their application gives possibilities for all kinds of deformations. One example of this are stock exchanges set up on the basis of trade law by active groups of people who created private organizations with very loose self regulation.

By the end of 1993 there were two stock exchanges in Sofia and more than ten stock exchange branches of the commodity exchanges. The legislation on securities and stock exchanges is still in preparation and that gives possibilities for many artificial exchange operations only to create an image of some new stock companies. The number of public stock companies is too small for so many exchanges, only about 30 companies offer their stocks to the public on the secondary market and their stocks may be bought on any of the mentioned stock exchanges. Control of the operations is very limited, if it exists at all on those exchanges.

Table 7.15
Market stages

Characteristics	Spontaneous markets	Establishment of new market institutions	Fully established markets
Legal framework	No modern legislation. No application of the old legislation.	Separate new laws No adequate legal system.	Legal system equal for all participants in the market. Market driven decisions.
Trade	Restrictions on trade are not applied and its violations are not punished.	Permanent introduction and change of restrictions with individual preferences.	Liberal trade system with universal preferences for everyone.
Prices	Speculative prices based on monopolistic position in payments.	Unstable prices, result of the tiny market, government intervention and speculation	Demand and supply driven equilibrium prices.
Access to markets	No regulation on market entrance and no rules for liquidation.	Dominating state sector. No bankruptcy in the state sector.	Free access to the market and equal position of private and state-owned enterprises.

Real markets may be developed only when the private sector dominates as an economic power. A large state sector, strong government and political interference, which is the Bulgarian case, still keep capital markets below real needs. But in some areas by the primary market of money deposits, the real process started with fair regulation under the pressure of real needs. In 1992 the increase of term deposits was 129 per cent compared to the previous year and in 1993, 85 per cent. They became the main element of the broad money with 60 per cent of the money supply in 1993. The main features of these stages are given in table 7.15. The process itself may be seen in the development of the bank sector and financial instruments.

Development and infrastructure of money and capital markets

Banks, exchange offices, exchanges: these are the most attractive and popular financial institutions of the entrepreneurial spirit of the businessmen in Bulgaria. Their development started right after the political crash of the old system with or without governmental interference. The banking system of the centrally planned economy was represented by four relatively separate institutions:

1   the Bulgarian National Bank (BNB) - the emitting institution which at the same time had direct contacts with the clients' enterprise and the population with more than 50 branch offices all over the country;

2   the Bulgarian Foreign Trade Bank (BFTB) - oriented entirely toward foreign trade operations and foreign exchange;

3   the Bulgarian Investment Bank (BIB) - which dealt only with investment projects including private construction of houses;

4   the State Saving Bank (SSB) - oriented mostly to arrange saving accounts for individuals through thousands of local offices.

This clear distinction and functional specialization was possible because of the limited role of money at that time. In 1987, by decree and without changing the banking law, seven commercial banks specialized in separate sectors of the economy were created. The role of the BNB changed slightly because it became a general depositor for commercial banks. This situation changed dramatically in 1990 when all branches of the BNB became separate banks. At that moment the creation of private banks without a pre-existing legal base started.

The First Private Bank (FPB), which is still one of the most popular new banks, was created as a stock company in 1990 with no governmental participation. Two former branches of the BNB in Kremikovtsi (near Sofia) and Petrich (near the Greek border) became private by enlarging their capital with participation of private firms. This was the irregular period of the banking system development. The approval of two acts was oriented directly to the banking system: the Bulgarian National Bank Act of June 1991 and Banks and Lending Act of March 1992 made possible the centralized policy for consolidation of banks which was one of the conditions for IMF participation in Bulgarian reform. The BNB, as a central bank, is independent from the government and subordinate to the parliament. This created the possibility for independence from political changes in monetary policy based on macro-economic analysis. The changes were significant:

1   Previous segmentation of banks (for example, before 1990, the BFTB was responsible for 94 per cent of the total foreign liabilities and the moratorium of March 1990 was announced not by the government but by the BFTB) was abolished and in all, more than 70 commercial banks were authorized to compete with each other. Nevertheless, they were allocated by territory and internally specialized by type of enterprise and the real competition was limited. Created in 1987 was a sector of commercial banks specialized, for example, in economic industry - the bank 'Electronics' has now no limitation in bank services.

2   Previous concentration, when in 1990 the three largest banks (BNB, BFTB and SSB) held more than 60 per cent of the total assets and deposits and more than 70 per cent of the total capital, is undergoing more equal distribution between seven to ten large banks, including private ones. At the end of 1993, the FPB signed an agreement with the South Korean company Daewoo for the enlargement of its capital to one billion leva which will make this bank the largest by equity capital after the BFTB; 39 per cent of the shares will belong to Daewoo. The list of Bulgarian banks ranked by total equities is given in table 7.16.

3   Branch concentration does not exist anymore. Most of the commercial banks had no branches in the country. They operated on limited territory and for different sectors of economy. The new situation is much more favorable for the creation of real competition in the money market in the near future with the presentation of several banks in all urban centers, which total more than 300 in the country. At the end of 1993 most of the commercial banks had between 15 and 50 bank offices.

4   Ownership; before 1990 all banks were state-owned. The wild period of no legal regulation gave possibilities for the creation of the first private banks and the privatization of some small commercial banks. The privatization process was made in an irregular way as it was declared in the statement of the BNB. The BNB has not given any

permission to enlarge capital and to sell stocks to private companies. But the fact is already accepted without further consequences.

5 Consolidations of banks created larger banks with possibilities to compete on the international market. The existing large number of banks in 1991 (over 70) was reduced to 12 in 1993 with still undecided futures for some of them. The consolidation scheme is given in table 7.17.

The Bank Consolidation Company was created in February 1992 with the mission not only to concentrate capital but to perform the process of privatization of banks. But a series of scandals about hidden privatization made possible the decision of the parliament in 1993 to postpone bank privatization by two more years. The Consolidation Company is the stockholder of all state-owned banks and only with its decision may the equity capital of banks be changed and sale of shares to strategic, mostly foreign, investors be made. When BCC fulfills that role and all the shares are sold, the company will be dissolved. Still this is not planned for the near future.

The new banking law, Banks and Lending Act, allows non-banking financial firms to provide security transactions, management of capital investment, etc. Thus, the new financial companies will develop their activities under the control of the BNB, without any state control and even against state intervention.

The main problem of the Bulgarian banking and financial systems are the so-called bad debts. They were created by the former totalitarian practice to plan investments without any relation to the effectiveness of the enterprises. In the situation of crisis in the transition period it became worse because of the lack of financial discipline of the state firms, recipients of credits in the absence of legal procedures for bankruptcy and interlinked interest of state-owned and private firms to return risks to the state.

At the end of 1992, the total amount of bad debts according to the BNB's annual report, was about 70 per cent of GDP. At the end of 1993 the parliament accepted in principal the law which transfers all bad debts established before the end of 1990 and not served before June 1993 to state banks with maturity of 20 years. This gives the banks a chance to overcome the risk of bankruptcy.

Table 7.16
Bulgarian commercial banks in 1992, ranked by total equity
(values in million leva)

Bank	Common stock	of which state-held	Reser-ves	Total equity	Total assets	After tax profit
			State-owned banks			
BFTB	1,200.00	1,179.70	3,658.62	4,858.62	72,500.00	1,114.80
Mineral Bank	291.78	130.05	3,183.45	3,475.23	27,547.00	-682.50
SSB	800.00	800.00	2,100.00	2,900.00	n.a.	n.a.
UBB[a]	679.47	298.00	368.70	1,048.17	25,965.90	0.02
Biochim Bank	160.00	80.59	358.78	518.78	8,610.54	20.43
Vazrazhdane Bank	26.69	15.90	486.94	513.63	873.17	0.87
Sofia Bank	200.00	200.00	153.31	353.31	5,467.72	301.07
Balkan Bank	200.00	117.69	152.53	352.53	14,383.50	0.00
Agrocooperative Bank	90.00	63.25	110.31	200.30	7,980.34	-4.38
Plovdiv	62.47	53.23	83.53	146.00	4,091.26	5.83
Elektronika Bank	68.35	65.12	58.35	126.70	4,958.43	-419.00
Transport Bank	47.40	46.46	78.93	126.33	3,203.37	74.42
Varna	36.50	36.01	84.68	121.18	2,032.09	40.50
Economic Bank	100.00	63.53	0.00	100.00	37,872.59	159.60
Bourgas	24.95	22.19	65.34	90.29	2,402.81	12.32
Blagoevgrad	18.00	17.04	66.29	84.29	1,577.03	5.39
Stara Zagora	35.43	33.84	38.97	74.40	1,791.97	3.59
Hemus Bank	35.60	23.99	34.77	70.37	1,005.12	9.28
Sredetz Bank	13.69	12.10	55.55	69.24	1,944.82	6.92
PostBank	60.00	60.00	9.14	69.14	2,543.73	23.66
SirBank	44.25	41.45	10.32	54.57	1,107.51	8.42
Veliko Turnovo	21.10	19.40	32.71	53.81	1,181.91	8.62
Razgrad	12.67	10.99	36.37	49.04	730.07	0.00

Table 7.16 (continued)
Bulgarian commercial banks in 1992, ranked by total equity
(values in million leva)

Bank	Common stock	of which state-held	Reser-ves	Total equity	Total assets	After tax profit
			State-owned banks			
Vidin	12.17	11.32	30.69	42.86	899.51	1.02
Kazanlak	21.00	20.22	14.21	35.21	1,045.96	0.00
Vitosha Bank	9.60	9.26	25.10	34.70	810.04	0.00
Kyustendil	18.40	18.19	13.53	31.93	393.85	6.03
Karlovo	21.79	21.03	6.58	28.37	1,039.34	-59.57
Trakia Bank	16.90	15.47	8.87	25.77	860.99	0.00
Troian	7.00	5.34	17.36	24.36	282.19	4.58
Rila Bank	7.00	5.72	16.37	23.37	532.76	0.00
Gorna Oriahovitsa	9.51	8.80	11.51	21.02	619.85	4.47
Smolian	7.00	4.39	12.27	19.27	574.18	4.95
Yambol	13.09	9.39	4.53	17.62	3,352.08	-244.70
Silistra	13.13	11.65	3.07	16.20	736.34	0.02
Asenovgrad	7.00	4.97	8.46	15.46	299.72	6.53
Cherven Briag	7.00	6.78	7.46	14.46	340.46	0.11
Provadia	7.00	5.49	6.03	13.03	279.47	-1.71
Devin	7.00	6.63	4.70	11.70	88.41	1.51
Parvomai	7.00	5.13	3.99	10.99	153.28	0.80
Elin Pelin	7.00	6.30	3.45	10.45	128.32	0.20
Liaskovets	7.00	6.15	1.85	8.85	141.76	0.72
Gotse Delchev	7.00	6.85	0.92	7.92	290.57	0.00
Chepelare	7.00	6.43	0.83	7.83	155.04	4.78
Mezdra	7.00	6.26	0.43	7.43	260.46	0.00
Subtotal	4,455	3,592	11,430	15,885	243,016	100

Table 7.16 (continued)
Bulgarian commercial banks in 1992, ranked by total equity
(values in million leva)

Bank	Common stock	of which state-held	Reserves	Total equity	Total assets	After tax profit
			Private banks			
FPB	337.01	0.00	7.43	344.44	8,406.43	79.70
CCB[b]	223.32	0.00	41.56	264.88	1,284.84	8.13
BAC[c]	150.00	63.35	109.61	259.61	6,354.50	271.81
FEIB[d]	200.00	3.15	30.38	230.38	1,825.51	8.64
Dobrich Tourist	30.00	3.49	139.42	169.42	2,007.15	44.81
Sport Bank	150.00	11.28	0.00	150.00	3,201.94	32.26
PAIB[e]	110.00	0.00	10.00	120.00	500.59	0.04
Credit Bank	61.02	0.00	0.00	61.20	1,647.40	2.19
Crystal Bank	20.42	4.93	18.29	38.71	406.74	17.17
Business Bank	27.55	4.66	3.88	31.43	402.28	8.68
Agro-business Bank	22.00	0.46	0.08	22.08	246.36	1.96
ITDB[f]	14.30	0.00	0.00	14.30	n.a.	n.a.
Subtotal	1,346	91	361	1,706	26,284	475
Total	5,801	3,684	11,790	17,591	269,299	576

a United Bulgarian Bank
b Central Cooperative Bank
c Bank for Agricultural Credit
d First East International Bank
e Private Agro Investment Bank
f International Trade and Development Bank
Source: BNB.

The first financial markets which appeared after 10 November 1989 were the money markets. At the beginning they were irregular markets. They started as black markets long before the political fall of the totalitarian state but they were very active during the start of the economic reform in February 1991. The irregular money market had its own exchange - a highly concentrated market on the street in the center of Sofia known as the 'Magura street exchange'. Quotations of this exchange were carried in some of the major newspapers (24 Hours) for more than one year. The administrative measures against the 'Magura street exchange' had no significant results. The central banking exchange rate became acceptable only after an unofficial agreement with the first free currency exchange 'Magura'.

The transformation of the irregular money market into a regulated market was accomplished with the governmental decision for free exchange of foreign currency in the banks and the newly established exchange offices. Since early 1993, more than 1,000 exchange offices are open all over the country, the majority of which are private. All this new infrastructure had to follow the central bank exchange rate which became the first most important and most popular financial information in the country. The introduction of the floating exchange rate and the intervention of the BNB only as one of the market makers was the first step in the creation of real financial markets. The intervention of the BNB was only about 11 per cent in 1993. This fact shows that the market is not artificially imposed but is based on demand and supply. Exchange between commercial banks only has 25 per cent of the market. Nevertheless, the exchange rate in 1992 and 1993 was very stable. For 1992 its growth was 14 per cent and for 1993, 16 per cent with some fluctuations in between. Speculation created a crisis in November 1993 when, for two days, the growth was 15 per cent followed by a fall which created a small increase in general tendency, so the devaluation of the leva to USD in 1993 was 25.3 per cent.

Analysis made by the State Agency for Programming and Development shows that the foreign exchange market does not react to fundamental factors which influence the exchange rate and in the first place to the position of the dollar in the world market. The exchange practice in Bulgaria is oriented to the dollar. All transactions are evaluated in dollars and on this basis the BNB announces the fixing rate which is only for accounting purposes. It is not obligatory, as in the Japanese model, but it is the orientation for all real transactions during the day. The exchange rate of other currencies is based on crossrates on the Austrian money market. This creates great possibilities for speculation through swaps to different currencies.

Table 7.17
Consolidation of Bulgarian state-owned commercial banks in 1992-93

Consolidated Bank	Incorporates following banks
Consolidation completed or under way	
United Bulgarian Bank	Doverie Bank, Construction Bank, Rousse, Pleven, Vratsa, Shoumen, Haskovo, Gabrovo, Kardjali, Lovech, Montana, Pazardzik, Peshtera, Sliven, Nova Zagora, Iskar, Elhovo, Botevgrad, Samokov, Targovishte, Pernik, Popovo
Balkan Bank	Balkan Bank, Vidin, Gorna Oriahovitsa, Liaskovets
Hebros Bank	Agrocooperative Bank, Plovdiv, Veliko Turnovo, Blagoevgrad, Vitosha Bank, Mezdra, Troian, Chepelare
Express Bank	Transport Bank, Varna, Vazrazhdane Bank, Kyustendil, Silistra, Razgrad Smolian, Cherven Briag, Rila Bank, Gotse Delchev, Provadia, Devin
Sofia Bank	Sofia Bank, Hemus Bank, Electronica Bank, Kazanlak

Table 7.17 (continued)
Consolidation of Bulgarian state-owned commercial banks in 1992-93

Consolidated Bank	Incorporates following banks
	Planned consolidations
Biochim Bank	Biochim Bank, Sredets Bank, Bourgas Stara Zagora, Karlovo, Trakia Bank, Asenovgrad, Parvomai, Elin Pelin
	Undecided
Bulgarian Foreign Trade Bank Mineral Bank Economic Bank Post Bank Sir Bank Yambol	

Source: Bank Consolidation Company.

The exchange rate is mostly influenced by the intervention of the BNB and the difference between the interest rate for deposits in leva and in foreign currency. That satisfactorily explains the stability of the exchange rate, the revalorization of the leva and the change of the total amount of deposits in the banking system.

An interesting phenomenon which was born of the lack of legislation and the desire to make profits from the new Bulgarian capitalistic institutions is the boom of bourses. In Bulgaria there are now more than forty commodity exchanges. The first exchange was created in June 1991, the Sofia Commodity Exchange, which is still the largest and most prominent one. All exchanges are private organizations. Some of them are not-for-profit organizations but the majority are limited companies which want to attract consumers by the creation of open markets. Very often they are hybrids between commercial and brokerage houses. There is a large diversity of membership requirements, constitution rules and dealing procedures in the existing bourses. Some of them have no organized floor, nor regular sessions. The participation fee varies from leva 500 ($ 20) to leva 50,000 ($ 2,000). Some of the houses have no authorized brokers to allow out of session contracts. Some of them function as daily wholesale markets for agricultural products. Both parties of a contract are charged from 0.3 per cent to one per cent commission. Brokerage allowances are negotiable.

The assortment of goods traded on bourses is unlimited. This fact is easily explained by the desire to attract more clients and thus to ensure higher commission for the bourse. There are no standard contracts. The financial guarantee for the contracts and transactions is not available and clearing houses are not established. This is the reason for not having forward and futures markets.

The stock exchanges are in a similar position. There are only two specialized stock exchanges and more than ten stock departments of commodity exchanges. About ten companies (mainly banks) offer shares on the secondary market. The primary market started in 1990 with the creation of the first public stock companies but until now it is very limited. The secondary market started in 1992 with the creation of the First Bulgarian Stock Exchange. The delay of the privatization process made the stock exchanges entirely artificial market places with no real customers and very small turnover. The main customers are the same banks and financial companies which are members of the exchanges. The average weekly quantity of exchanged shares is 250,000 for the last year on the main exchange, the First Bulgarian Stock Exchange. This amount is certainly not enough for the exchange to exist on the commission on the trade. The exchange is artificially maintained in order to create some traditions while waiting for better times.

The experience of the state stock exchange as an institution had no great success. It was founded in 1915 under the law of 1907 and existed until 1940 with two interruptions of two to three years because of the lack of interest. In the new conditions stock exchanges can work effectively in the privatization process with techniques including sale of shares of the privatized companies only through commodity exchanges. In this situation, laws regulating the stock exchanges will be necessary. It is interesting to notice that at the beginning of the century the law was introduced much earlier than the real business activity. In the period of transition business action comes before the legislation. That creates possibilities for the existence of irregularities.

Financial instruments development

The most popular instrument was money deposits in commercial banks, which was used not only by households but also by firms for temporary free money, starting from 1991 when prices were liberated and interest rates jumped dramatically from one to three per cent to over 60 per cent on an annual basis. Deposits are the only instrument which has acceptable liquidity and relatively low risk giving fair income. Nevertheless, the real interest rate on deposits for both 1992 and 1993 was negative. In 1992 the CPI was 79.5 per cent and the average interest rate on deposits around 54 per cent. In 1993 the CPI was 63.4 per cent and the average interest rate was 52 per cent. According to the new budget proposal, in 1994 the expected CPI of 40-45 per cent will mean positive interest rates for the first time. But in May 1994 inflation was already 42 per cent. But many depositors are turning from leva to dollar deposits. In that case deposits may remain the major instrument of hedging for households. In 1992 the average growth of deposits on a quarterly basis was 10 per cent and in 1993, 7 per cent. Lack of well-developed secondary markets on securities and understanding of the securities market by investors still gives priority to this instrument. But in the situation of insufficient foreign investments, household savings becomes the only net creditor in the national economy. At the end of 1993 the total amount of deposits was leva 108 billion which, compared to the expected 1993 GDP, is 40 per cent. This creates internal economic stability. The only negative side in this scheme is the use of savings for crediting unprofitable enterprises and the state budget.

New Treasury bills were also developed. The high inflation rate and high basic interest rate of the National Bank made the issue of obligations by firms insufficient. The attractive point of the Treasury bills was the possibility to use them for Lombard credits and to reduce taxes. The Lombard credits were open to the commercial banks from the beginning of 1992. The Treasury bills were also only available for commercial banks at that time. There was no possibility for physical persons or even firms to buy TBs as it was, for example, in Hungary since 1988. That is why in 1992 the interest in the auctions held by the BNB between the commercial banks was very limited. The change came at the beginning of 1993 when banks received permission to act as brokers for state and private firms in the auctions. At the end of the year bids received were about 20 times higher than bids starting in mid January. The demand was larger for three-month TBs (issued for the first time) because of the uncertainty in the basic interest rate and the expected high rate of inflation for 1993. The possibility of Lombard credits for firms made the TBs attractive. The market for Treasury bills is moving slowly from the artificially limited circle of the interbank market to the real market of bills, notes and bonds.

Secondary markets of Treasury bills started in September 1993 on existing stock exchanges and off the counter in financial brokerage firms which were active mostly in the sale of stocks. Two year state bonds (obligations) were offered with little success in the middle of 1993. Transfer of bad debts to long-term state securities might help to move from this artificially imposed market to a real market in state securities.

Stocks were introduced under Decree 56 in 1989 when the first joint-stock company was founded, Bulsovinvest. But the open market for stocks started only in 1992 with the creation of private banks and the start of the new non-financial commercial joint stock companies like Lex. In 1993 the new hit became investment funds. At the end of 1993 the stock market was operating with stock of 12 banks (including state-owned ones like Balkan Bank), 5 insurance companies, 6 investment funds and 6 non-financial companies. The weekly turnover is symbolic, 10,000-15,000 stocks on average at all ten stock exchanges. The usual value of stocks is leva 100 which means that the average weekly turnover is about leva 1.5-2 million.

This artificially imposed market which was initiated not by the government but by entrepreneurial organizations may turn into a real market with the approval of the privatization law by the national assembly. It introduces in Bulgaria the mass privatization which is popular in most of the other East European countries.

Main features of this new legislation are:

1   the right of any Bulgarian citizen over 18 years old to receive a privatization check for leva 25,000;

2   the preliminary payment to obtain the check is leva 900 or 3.6 per cent of its value;

3   privatization checks are valid only to obtain stocks of enterprises and investment funds for privatization;

4   500 enterprises will be announced in a special list for privatization and ten state controlled privatization funds will be established;

5   this system does not replace the right of personnel to obtain under a preferential regime up to 20 per cent of the total of the enterprise they work in;

6   special time tables give terms for transfer of checks into shares of desired companies or funds.

After that period a boom of the secondary market is expected.

Main factors for the creation of real capital markets are:

1   modern legislative system;

2   effective governmental regulation;

3   creation of a critical mass of actors on the market for each financial instrument;

4   permission for speculation on financial markets.

The slow privatization process reflects the slow development of financial markets in Bulgaria.

Conclusion

The main conclusions of this study are as follows:

1   The delay of the reforms in Bulgaria is mainly a result of political instability and internal controversy which paralyzed adoption and application of new laws. Priority given to the restitution and political biases against mass privatization created an atmosphere of suspiciousness of any privatization act. None of the four different governments supported by different political forces could overcome the privatization complex. Only a normal democratic atmosphere can provide a positive background for effective transition to a competitive market.

2   The development of competitive money and capital markets passes in different ways three main stages: spontaneous markets with possibilities for rapid concentration of capital; the imposition of market institutions with a strong regulative role of the government; and, finally, creation of competitive centers - market makers oriented mostly around commercial banks. This formula is a natural result of the development of market focus from the individual entrepreneurial activities to the institutionalized effective market regulation.

3   International support and use of already existing models for the development of financial markets is of crucial importance for the success of the transition process. The IMF and the World Bank had a very stimulating role in the financial stabilization of the country and the development of the financial sector, not only with financial support, but also with the know-how and, what is of great importance, control on implementation of intents. The competition of different models of financial markets, American, Continental and Japanese, introduced by different groups, formed a specific mixture of different approaches from different models. The dominance of one model will probably depend on how rapid Bulgaria's association to the European Union will be.

4   The rapid development of financial markets without the support of the equal development of industry and private property creates a potential

223

instability in this market. There are many indications for the possibility of failure of both money and financial capital markets. Governmental regulative policy will be of essential importance for the success of these new markets.

5   The main source of the development of the money and capital markets is the entrepreneurial spirit of a large part of the population. In spite of a lack of theoretical knowledge and practical experience, the wish and sometimes curiosity to start a new business as it is in developed countries plays an important role in the introduction of new market activities. A strong incentive for that is the possibility of gaining personal wealth in a very short time. That is the reason for the reorientation from trade to financial markets of many advanced private firms and the creation of holdings with financial and brokerage houses. The novelty attracts.

Notes

1. Center for the Study of Democracy (1993), *The investment conditions in Bulgaria: the point of view of foreign investors. Results of case-studies of entities with foreign participation*, Sofia.
2. See *Successfully managing investments in Bulgaria. A survey*, Deloitte Touche Tohmatsu International, 1992.
3. *168 Hours*, no. 27, 1992.
4. The successful story of privatization - 'Zarevichny Producty'. The Belgium firm 'Amylum', which competed with two other foreign investors with the mediation of Austrian and British consulting firms, signed a contract for 80 per cent of the share of the Bulgarian firm 'Zarevichny Producty' for food processing of corn. The total amount of the contract is $ 45 million; 80 per cent is provided by the newly consolidated United Bulgarian Bank (UBB) and one per cent is for the personnel. 'Amylum' gave to the state the amount of $ 528 million and is obliged to invest in a two year period not less than $ 20 million and to preserve the existing working places. The process of negotiation was dramatic because at the last moment the third negotiator - the commercial bank 'Biochim' did not accept the debt swap; ownership and UBB had to pay to 'Biochim' leva 193 million for the debts of the enterprise.

References

Abel, I., Bonin, I.P. (1993), 'Capital markets in Eastern Europe: the financial black hole', *Journal of International Law*, no. 3.
Agency for Economic Coordination and Development (AECD) (1993), *Bulgarian economy in 1992*, Sofia.
AECD, (1993), *Bulgarian economy in 1993* (Annual report), Sofia.

Asselain, I.C. (1991), 'Convertibility and economic transformation', *European Economy*, no. 2, special edition.

Brainard, L.I. (1991), 'Reform in Eastern Europe: creating a capital market', *Economic Review*, no. 8, p. 49.

Dobrinsky, R. (1993), *Economic outlook for Bulgaria 1994-98*, Paper for the project LINK, Beijing.

Dobrinsky, R. and Grosser, I. (1993), *Bulgaria 1993: country report risk assessment*, WIIW, Vienna.

Guenov, K., (1993), *The monetary policy in 1992: Instruments and results*, AECD, Sofia

Hrncír, M., Klacek, J. (1991), 'Stabilization policies and currency convertibility in Czechoslovakia', *European Economy*, no. 2, special edition.

Hughes, G. (1991), 'Foreign exchange, prices and economic activity in Bulgaria', *European Economy*, no. 2, special edition.

Jackson, M. (1994), *Privatization, company management and financial development in Central and Eastern Europe*, this volume.

Jackson, M. (1991), 'Constraints on systemic transformation and their policy implications', *Oxford Review of Economic Policy*, vol. 7, no. 4.

Kornai, I. (1990), *The road to a free economy shifting from a socialist economy*, WW Norton, New York.

Miller, J. (1993), *The price index gap: a window to understanding Bulgarian economy*, Bulgarian Academy of Sciences (BAS), Sofia.

Murphy, A. and Sabov, Z. (1992), 'Empirical analysis of pricing efficiency in the Hungarian capital markets', *Applied Financial Economics*, no. 2, pp. 63-78.

Panov, O. (1991), 'Radical changes in the economics of socialist countries: a driving force for East-West cooperation' in Razvigorova and Wolf (eds), *East-West joint ventures: the new business environment*, Blackwell, London.

Thorno Alfredo (1992), 'Issues in reforming financial systems in Eastern Europe: the case of Bulgaria', *Working Papers Series WPS 882*, The World Bank, Washington.

8 Impediments to financial restructuring of Hungarian enterprises

Istvan Abel and Kristofer Prander

Introduction

The adoption of a market economy in Hungary, and in other East and Central European countries, has led to the emergence and growth of a large number of firms with risky future prospects. These events have also had an influence on the banking sector's own expectations of success, as well as the possibility of raising business taxes to reduce the Government's budget deficits.

These problems all originate from the legacies of the old system, when the state borrowed substantial amounts from abroad to finance a relatively high living standard in Hungary. Credits or grants were then extended to state enterprises, not in accordance with their relative efficiency, but in a more unstructured manner. Today, when there is high inflation, and the sources of new financing are rather limited, this leads to the effect of punishing the companies with the largest inherited debt instead of the ones that are the most inefficient. This paper argues that without adequate additional financing, i.e. as a consequence of a tight monetary policy, the problems of indebted enterprises are aggravated. In addition to this the large amount of bad debts inherited by the banks when they were spun-off from the National Bank of Hungary made it impossible for them to assist the companies in an efficient way, especially since the Government's large foreign debt and budget deficits has made it necessary for the Government to take possession of the major part of Hungary's savings.

A Ministry of Finance survey of the tax returns of 57,200 companies reveals the magnitude of the loss-making enterprises in Hungary.[1] Nearly half of all Hungarian enterprises are loss-making. These enterprises hold 59 per cent of all bank debt and employ 1.1 million people. Nearly two-thirds of these losses are concentrated in 603 firms with HUF 243 billion in outstanding debt (or 35 per cent of all enterprise loans) employing

227

400,000 people. According to the survey the operating cash-flow of these loss makers, after subtracting for depreciation, amounted to negative HUF 317 billion or 11 per cent of GDP. The cash needed to make up for this enormous operating cash-flow deficiency could have been raised through selling assets, extending inter-enterprise credits, leaving taxes unpaid or through increased loans from the banks.

The figures revealed by this study make the financial restructuring of these companies a very urgent issue. This in turn places high requirements on the developments of the Hungarian financial sector and the monetary policy pursued.

A financial system in a market economy needs to fulfill four main objectives. First, its banks must safely and efficiently facilitate payments, which is the core of any banking system. Second, its financial institutions should mobilize savings and channel them to the most efficient uses. Effective intermediation means a larger response to reforms, whereas the misallocation of resources can undermine the very structural reforms that a country adopts by stifling competitive firms and supporting the uncompetitive. Third, a financial system should offer firms the means to package, price and diversify risk. Fourth, the banking sector in particular may have a role in exercising control over enterprises either indirectly through its credit policies, or directly through its decision to pursue bankruptcy proceedings, its involvement in company boards, or as a partial owner.

In addition to this, firms have four sources of finance outside their own funds. First, they can rely on funds from the Government and the state's industrial policy. This is the approach taken in former East Germany by the Treuhandanstalt. Alternatively, firms can turn to the capital markets. The Budapest Stock Exchange was established in 1990 but it started to take off only in early 1994 not giving much of a chance to solve the financial problems of mostly financially weak enterprises. A third option is inter-enterprise credit. This option, however, can be problematic for many reasons. Fourth, the firms can turn to banks to obtain loans.

This chapter begins by outlining the major factors constraining the development of the Hungarian financial sector. This includes the extremely high need for financing in the general Government and the weak capital position of the banks. In addition to that we mention the relationship between trade credit and inflation for an individual firm. While the implications of the relationship are well-known, they seem to have been largely forgotten by policy makers who have adopted monetary policies consonant with those in western industrialized countries. We then demonstrate that the type of restrictive policy pursued by the central banks plays a critical role but one which is likely to reduce the chances of a successful transition to a market economy. This is the subject of section four. Section five deals with other issues related to the debts accumulated by firms, and the implications for the firm of constraints on enterprise liquidity.

Constraints on the development of the Hungarian financial sector

General government

In 1992 the Government showed a deficit of 7.4 per cent of the GDP and in 1993 the estimate is that the deficit will equal HUF 200 billion or 6.6 per cent of the GDP. The main contributing factor to the deficits is that the decline in total performance of the economy has been steeper than expected, thus reducing the tax revenues. In addition, the high levels of unemployment have increased the social security and pension expenditures of the state. Finally, the large foreign debt inherited from the former regime requires considerable debt service.

These factors have meant that the state is left with a large need for borrowed funds. This has been accomplished partly by entrance into foreign capital markets, but mainly by the absorption of the country's savings. Since the growth of household savings has not been able to keep up with the borrowing requirements of the state, the Government has issued bonds and Treasury bills to such an extent that it has squeezed out the liquidity of the market and raised interest rates, which increased the interest expenditures of the banks and their difficulties in obtaining funds.

Budapest Stock Exchange

On the Budapest Stock Exchange the trading in equity has been limited. In 1993, 69 per cent of the turnover on the exchange was made up by government bonds and Treasury bills. Thus the Government has crowded out the trading here too. Other factors contributing to the severely limited nature of possibilities for companies to obtain funds through the exchange is the limited historical financial data available on mostly new or reorganized companies, the recession, the uncertain and risky future prospects of many companies and the time-consuming nature of the procedures as well as listing requirements. However, it should be noted that it is typical even in a market economy, that when companies need funds the most, their access to capital markets is limited due to economic conditions; when interest rates are high making loans an inefficient source of finance, share prices are normally low, making it inefficient to obtain capital here. The restructuring of funds and the repayment of expensive bank credit is best done when the stock market is bullish, i.e. when interest rates usually follow a downward trend.

When the commercial banks were spun-off from the National Bank, they inherited large amounts of bad debts. Since these bad debts were not immediately eliminated by the Government, the situation worsened due to the recession and due to the almost simultaneous introduction of three very important laws. The New Bankruptcy Law, although taking effect legally on 1 January 1992, began to have significant impact on 8 April 1992 when a company with any outstanding debt more than 90 days overdue was required to initiate bankruptcy proceedings or the responsible parties would be subject to criminal prosecution. The New Banking Act, effective on 1 December 1991, introduced specific categories for rating the loan portfolios of the banks, mandated the accumulation of provisions against the qualified loans in the portfolio and specified a schedule for meeting capital adequacy targets based on these components. The Act on Accounting, effective 1 January 1992, mandated that companies revise balance sheets and other business documents to bring them closer to international standards.

Since regulators had not foreseen the large amount of companies filing for bankruptcy (both due to the strict bankruptcy rules and the tougher accounting principles), the administration was unable to handle them, thus leaving companies that could have been restructured in a situation where their value faded away, leaving the banks with larger amounts of bad debts. Value fades away since during the bankruptcy procedures, assets cannot be used and they cannot be sold. These 'queues' are also the reason why bankruptcy procedures are not a good way in which to handle restructuring. These facts have had the consequences that the banks were forced to reduce their risks which have had meant major implications for the Hungarian companies and the Hungarian financial sector.

First, the banks have preferred to lend to the Government at a lower risk than extending risky loans to corporations. Due to the Government's financial needs this has been very easy. Secondly, the banks had to increase their spreads (also in part due to implicit taxes - like high reserve requirements - by the National Bank of Hungary) and increase their short-term credit, instead of long-term borrowing. Also the investment loans have been decreasing. Thirdly, the banks have been less able to devote money to develop their technical standards and other important issues, needed to make financial intermediation efficient. This problem has two sides; first since a lot of money must be put away as provisions for bad debts, less money can be spent on technical development, thus delaying transfer times etc.; second, because money must be invested in technical developments, less can be used to limit the risks, and therefore contributes to keep up the spreads. Fourthly, it has meant that the healthy companies paid the costs of the damage caused by the poor situation of the rest of the economy. Since the banks are forced to increase their spreads to cope with the risks and accumulate provisions, they are left with the less attractive clients, since the healthy clients may get cheaper financing from joint venture banks having access to cheaper funding. As the supply of these funds is limited, those good clients who remain with domestic banks also face relatively high spreads, and so contribute to provisioning for the bad clients' position. Those companies which were able, either joint ventures that can obtain finance from their mother company in their home country, or foreign companies operating in the country using foreign banks that have not inherited a large amount of bad debts, sought foreign sources of finance. For the remaining corporations this means that since they cannot find sufficient financing from the banks they have to look for other sources or other ways in which to solve their financial needs.

Loan consolidation programs

The Government made two attempts to clean up the banks' bad portfolios. In the first phase of the loan consolidation scheme, the Government accepted HUF 102.5 billion bad claims that the banks sold. In exchange the banks received consolidation bonds, with a maturity of 20 years, of a value of HUF 79.4 billion. The banks had to write off the difference of HUF 23.1 billion. The problem was that this consolidation program did not aim at the problem itself, but only worked as short-term medicine. While it decreased the amount of bad loans in the short run, it did not remove the reasons why the bad loans occurred in the first place.[2]

Recently, the Government launched a new loan consolidation program, which involves eight banks. By December 1993 their capital adequacy ratio is going to be brought up to zero per cent and in May 1994 to four per cent. This program has been criticized because it does not reflect the banks' previous efforts to improve their bad debt situation. The banks, which had adopted hard policies, now lose to those which have just been concentrating on increasing their market share. Furthermore, the program includes a large amount of money, HUF 144 billion, and does not guarantee that the worst banks will not again fall into the trap. Another point of argument is that the Ministry of Finance increased its ownership share in the banks. One of the banks' largest problems earlier was that the state has been the controlling owner. This has created the wrong incentives for the banks.

Another option would be for the state to concentrate on a few banks, close to privatization. This strategy would have the advantage of a smaller amount spent by the state in the first round and the possibility of the state getting some of the money back when the banks actually are privatized. Furthermore the privatized banks would be more able to move the economy in the right direction.

Other financial intermediaries

Regarding the institutional investments that are important contributors to the financial markets in western countries, they are not of the necessary size yet. The insurance companies also invest in government papers, the state social security fund and pension fund go to cover these posts, and the investment funds' investments have been characterized by low liquidity due to tax advantages if they are kept over three years.[3]

The derivatives markets face problems in developing due to the illiquidity in the securities market, the lack of market makers and a fixed rate of interest. This leads to the situation that buyers and sellers can scarcely be found at the same time agreeing to the same conditions. This is also the reason that no firm is capital intensive and risk-taking enough to work as a market maker. Due to very high volatility the valuation is difficult and risky. A further problem is constituted by the lack of an appropriate fixed rate of interest to use.

Inflation and firm financing

The facts accounted for in the previous section (such as the Government's contribution in raising interest rates), the crisis in agriculture and exchange rate adjustments, include the largest reasons for inflation. One consequence of inflation is a reduction in the real value of the stock of outstanding debt. For firms which are indebted, this, of course, is good news. But as inflation rises, so does the nominal interest rate which can also lead to serious financial difficulties.[4] Consider a firm that is supposed to have inflowing funds of 200 and outflowing funds, including interest payments of 20, of 160 and debts of 400 at a real interest rate of 5 per cent when inflation is zero. This leaves the company with a positive cash-flow of 40. Now, let inflation rise to an annual rate of 30 per cent. Assuming the same real interest rate, nominal interest rates would have to rise to 35 per cent in order to leave lenders with the same real interest rate.[5] Also assume that all prices and outflows rise by 30 per cent as a result of accommodating increases in the money supply.[6] The nominal interest cost of the debt would thus rise to 140 (400 x 35). Thus, under the condition that access to new credit, such as during a credit crunch, is denied,[7] the firm with an otherwise sound cash-flow would now experience a negative cash-flow of 62 as a result of inflation. Only firms with zero debt to begin with, who would have the same positive cash-flow in real terms as before, or ones who reduce their debt to try to maintain the same level of cash-flow in real terms as before, would be unaffected by such a policy.

In 1990, the producer price index equalled around 18 per cent and the short-term credit rates were around 25-30 per cent. In the beginning of 1991, the producer price index (PPI) rose to 40 per cent, while short-term rates only increased a little, creating negative real interest rates. During

1991, the PPI decreased steadily to 6.9 per cent. As a consequence, real interest rates (i.e. nominal interest rates minus PPI) went up significantly to a level of almost 20 per cent. Lately, at the end of 1992 and the beginning of 1993, the gap has narrowed, since the short-term credit rates have been decreasing to around 25 per cent and the PPI increased to around 12 per cent which still meant a 13 per cent real interest burden which is extremely high. What makes things so difficult, is that there is such a big difference between PPI and CPI. Deposit and lending rates follow the changes in CPI, while the interest burden for companies is determined by their selling power influenced by PPI.

Figure 8.1
Interest rates on loan and producer price index

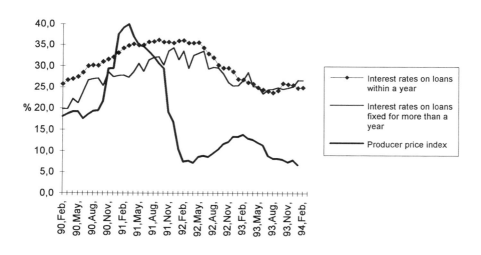

Source: National Bank of Hungary, *Monthly Report,* various issues.

Thus inflation has created a loss for firms unable or unwilling to increase their debt levels, necessitating a search for more debt to finance their loss thereby further exacerbating the problem of debt into a vicious circle as in the numerical example discussed above. Presumably, the policy also intends to raise, albeit perhaps only temporarily, the real interest cost of the debt to persuade firms, individuals and foreigners that the new monetary policy will be immediately reflected in nominal interest rates. During the 1980s, industrialized countries used precisely a high real interest rate policy to reduce inflationary expectations. It has been said, however, that the length of time necessary to achieve the credit constraining results is longer than in the past because well-functioning capital markets have produced a large number of derivative financial products to overcome such constraints at least temporarily (Siklos, 1993a). Such derivative products enable the necessary financial adjustments to take place without perhaps the large adjustment costs which would otherwise be expected. The difficulty in Central and Eastern Europe is that well-functioning capital markets are nonexistent, and the Government is crowding out private borrowers, as seen above. Therefore the trade credit continues to fulfill an essential function, but some firms will not be able to sustain higher debt levels for long before going bankrupt.

Monetary policy and the role of interest rates

It is perhaps not surprising that the worldwide emphasis on the control of inflation, originating in the stagflation of the 1970s, would carry over to the newly independent central banks.[8] The overriding concern with inflation manifests itself as a restrictive monetary policy. Restrictive here means high real interest rates to stem credit growth.

In order to understand the role of monetary policy as it influences economic activity let us consider the following illustration. Suppose the monetary authorities implement a restrictive monetary policy. Abel and Siklos (1993) suggest that firms can rely on trade credit to frustrate policy makers' attempts at reducing expenditures by the private sector.

One possible avenue is a quantity adjustment. Simply put, the quantity of trade credit in the private sector rises sufficiently to offset any restrictive monetary policy. Another possibility is an interest rate effect. In this case, firms reallocate their financial portfolio to offset or neutralize the impact of monetary policy.

Thus, one reaction to the shortage of credit would be for firms to ensure that the wage bill, among other costs under its control, be reduced. If this is the principal impact on all firms of the restrictive policy then national income would eventually be reduced via a reduction in expenditures.

Alternatively, firms could redirect some of their resources towards holding more financial instruments since these have a higher return than available alternatives. To the extent that higher yields are used to maintain the current wage bill there need not be any impact on labor income. Otherwise the tight monetary policy will have an indirect effect on incomes.

The objective of any restrictive monetary policy is, of course, to influence economic activity whether or not this is accomplished directly or indirectly. Its influence is most felt if all firms adjust in such a way as to leave the labor force unaffected and by reducing other expenditures such as the reduction of purchases, or through inventory adjustment.

Alternatively, the full weight of the adjustment is made on the labor force while the remainder of the firm's portfolio is unaffected. It may also be pointed out that a restrictive policy via interest rate changes may only have temporary effects.[9] It is worthwhile to enquire how the restrictive monetary policy in the transition to a market economy can be broken down into its output and interest rate components. However, this is an empirical question which we cannot deal with here but which is central to the problem of economic adjustment in the transition phase. Yet, it is important to understand these questions as they are critical to the success of the present monetary policy.

Let us now turn to the impact of current monetary policy on inflation in the former Eastern bloc countries. While the objective of a restrictive policy is to reduce inflation it is conceivable that inflation will follow from higher interest rates. One advocate of the latter possibility is Laffer (1970) who suggested that the definition of money should be altered to include a component for both the utilized and unutilized portions of trade credit.

Since trade credit represents purchasing power it should be reflected in any monetary policy decision meant to control spending and, therefore, inflation.[10] Thus, any policy meant to control an M1 type definition of the money supply will amount to nothing if trade credit is unchecked. In addition, of course, there is the question of the behavior of velocity. As an economy becomes more financially sophisticated velocity tends to rise secularly (Siklos, 1993a). Thus, in a quantity theory type setting, a reduction in money growth can easily be offset by growing velocity, thereby reducing the effectiveness of a high interest rate policy. Unfortunately, the literature on the role of trade credit in monetary policy is diverse and the empirical evidence inconclusive or contradictory.

Trade credit: the Hungarian experience[11]

Another study, conducted by the Hungarian Chamber of Commerce and the Economic Research RT,[12] seems to confirm the earlier mentioned survey made by the Ministry of Finance. In this, 8,000 businesses were asked about the conditions of their operations in September 1993. They received 1,200 answers which they claim reflect the Hungarian economic structure both in terms of industry and ownership. They see the low domestic demand as their largest problem, but the second difficulty they state is capital shortage. Two-thirds of the companies said they were short of capital and 21 per cent had a major shortage of capital. Many of the companies, 60 per cent, also report liquidity problems. They also have problems raising finance and with creditors' delayed payments. Seventy per cent state that financial shortages constitute a problem for growth. The main ways in which they obtain short-term capital is by overdrafting (50 per cent), cutting development budgets (30 per cent) and selling their assets (25 per cent).

This is also consistent with an earlier report by the Hungarian Institute for Economic Research, outlining the financing arrangements of enterprises (Gazdasagkutato Intezet, 1991). The study, based on 4,405 firms in 1988, reveal that the companies had serious liquidity problems and that the firms with the lowest liquidity ratios were also the ones for which receivables were substantially smaller than payables. It is also noteworthy that, as the liquidity ratio rises, receivables display a tendency to fall. This is explained by the fact that as liquidity concerns increase, trade credit rises in a parallel fashion. The correlation between receivables and payables was also found to be positive, but payables rise faster and receivables more slowly the worse the liquidity position of the firm. Thus profitability and liquidity, as measured in this study, appear to be positively correlated.

In addition to this, it was found in yet another study (Abel and Siklos, 1993, table 2), that total inter-enterprise debt rose 1,300 per cent between 1987 and 1992, while the number of enterprises engaged in such transactions also rose by over 1,300 per cent.

The number of firms with debt of HUF 25 million or more kept rising to reach a peak of 1,157 (See Abel and Siklos, 1993) in February 1992. The three largest commercial banks consider that 90 per cent of this debt is unbacked.

The introduction of the New Bankruptcy Law has conceivably had two effects which may give a misleading picture of the credit stance in the post January 1992 period. First, the number of firms reporting inter-enterprise credits fell. Second, all firms became more reticent in extending such credits since the risks associated with any losses are now significantly higher. While some argue that trade credit is a free source of funds, evidence suggests otherwise. Buyer costs; loss of cash discount - if the buyer would have received a cash discount, the trade credit is sure to be a very expensive source of finance. Higher risk of illiquidity leads to higher cost of finance from other sources. Seller costs and increasing collection periods are all additional expenses of financing a larger size of receivables.

The consequences of the firms overdrafting or rolling-over of loans is that this places the banks in a more difficult position by increasing the bad debts and in the case where credit is extended to inefficient firms, leads to a situation with a bankrupt company.

Instead of using capital for investments that could ensure some growth in the economy, capital is used to fulfill the Government's financing need for unemployment compensation, social security contributions and to cover interest expenses on debts, that is, to finance consumption and not investments.

Conclusion

About half of the Hungarian companies are loss-makers. The main cause of the losses of course is the recession but the large debt inherited from the old system is also an important contributing factor to the problem. The Government's large financial needs have created a credit crunch and crowded out enterprises from the lending markets. Therefore the financial sector has not been able to meet the demand created by the extremely concentrated need for financing in the first period of the transition toward a market economy.

We have argued that the pursued monetary policy has not helped to ease the tensions between the supply and demand for financing. Instead, the high interest policy has been circumvented by an increase in trade credit and the velocity of money circulation, which in addition has contributed to fuel inflation. There is an urgent need for changes to improve the conditions of growth. To reach these goals we recommend the following:

1   Concentration should be focused on reducing the budget deficit as a share of the GDP. This should be achieved by changing the current fiscal policy. At the same time to improve the supply of funds, a reduction in the implicit taxation of the banks is desirable. A first step in this direction is to reduce the high reserve requirement rate currently imposed by the National Bank of Hungary.

2   Those banks which are closest to privatization should be privatized as soon as possible. This would bring them the needed capital, access to cheaper funds, better technology and improved governance. These banks would then be able to lead the development of the Hungarian financial sector. This is best achieved by foreign investors, since a large enough investor is nonexistent in Hungary today. Also, a foreign investor would be able to bring the banks the needed improvements listed above. To achieve this target, it is necessary for the Government to make a policy decision, since a foreign investor probably would require controlling interest.

State banking supervision should be improved by more independence, larger competence and better coordination.

Notes

1. Ministry of Finance (1993).
2. Compare the results of Hrncír (1993).
3. Investment funds in 1993 in Hungary included: Creditanstalt, Hunnia I, Hunnia II, Horizont, Budapest I, Budapest II, Buda, Buda State Sec., Premium, Bonus, Profit, Millenium, Alpok, Taller, Arany Taller, OTP I, Piller I, Piller II, Patex, Hungary 2000.
4. See also Li and Pradham (1990) who base their results on the theory outlined in Wadwani (1986). The problem of inflation and trade credit are also considered in Gordon (1982) and National Bank of Hungary (1993).
5. This is the result of the well known Fisher effect. The result is expected in equilibrium which, in financial markets at least, is likely to be attained quickly, especially as inflation accelerates.
6. This too is an equilibrium result best understood in terms of a quantity theory type relationship.
7. See Shaffer (1992).
8. Independent in the sense of legal or statutory independence. For a survey of the literature on central bank independence in Central Europe see Siklos (1993c).
9. If interest rates are changed infrequently. Alternatively, frequent interest rate changes could also lead to temporary effects because the costs of reacting to each one of them rise with the frequency of interest rate changes. In other words, interest rate volatility can make it difficult to distinguish temporary from permanent effects of interest rate changes.
10. See also the conclusions in Hrncír (1991).
11. The Hungarian experience seems to be one for which there is relatively sufficient information to conduct an analysis.
12. GKI (1993).

References

Abel, I. and Siklos, P.L. (1993), 'Constraints on enterprise liquidity and its impact on the monetary sector in formerly centrally planned economies', *CEPR Discussion Paper*, no. 841.

Gazdaságkutató Intézet (Economic Research Ltd.) (1990), 'A magyar vallalatok finanszirozási szerkezetének fobb jellegzetességi' (Characteristics of the structure of financing Hungarian enterprises), *GKI Tanulmányok*, Budapest.

GKI Gazdasag Kutato (1993), *A monetaris szfera varhato fejlodese 1993 masodik feleben*, Budapest.

Gordon, R. H. (1982), 'Interest rates, inflation and corporate financial policy', *Brookings Papers on Economic Activity*, no. 12, pp. 461-491.

Hrncír, M. (1991), 'Money and credit in the transition of the Czechoslovak economy in Siebert, H. (ed.) (1991), *The transformation of socialist economies*, J.C.B. Mohr, Tübingen.

Hrncír, M. (1993), *Reform of the banking sector in the Czech Republic*, paper presented at the conference on Development and Reform of the Financial System in Central and Eastern Europe, Vienna.

Laffer, A.B. (1970), 'Trade credit and the money market', *Journal of Political Economy*, vol. 78, March/April, pp. 239-267.

Li, C. and Pradhan, M. (1990), 'Inflation, financial liberalization and bankruptcies in Argentina' in Phylaktis, K. and Pradhan, M. (eds), *International Finance in Less Developed Countries*, St. Martin's Press, New York., pp. 98-113.

Méro, K. (1990), *Vállalatfinanszírozás Magyarországon*, *Gazdaságkutató Intézet*, Budapest, pp. 38-151.

Ministry of Finance (1993), *Tasks of the implementation of the bank and debtor consolidation*, Budapest.

National Bank of Hungary (1993), *Monthly Report*, no. 8-9, Budapest.

Shaffer, M.E. (1992), 'The Polish state-owned enterprise sector and the recession in 1990', *Comparative Economic Studies*, no. 34.

Siklos, P.L. (1993a), 'Income velocity and institutional change: some new time series evidence: 1870-86', *Journal of Money, Credit and Banking*, no. 26.

Siklos, P.L. (1993b), *Money, banking and financial institution: Canada in the global environment*, McGraw-Hill Ryerson Ltd., Toronto.

Siklos, P.L. (1993c), *Central bank independence in Central Europe: a preliminary investigation*, paper presented at the conference on Development and Reform of the Financial System in Central and Eastern Europe, Vienna.

Wadwani, S.B. (1986), 'Inflation, bankruptcy, default premia and the stock market', *The Economic Journal*, no. 96, pp. 120-138.

9   Privatization and capital markets

András Giday and Agnes Sári-Simkó

Introduction

The rebuilding of the market economy requires both privatization and the development of capital markets. A well functioning capital market is able to correct privatization. Hungarian privatization rests on 'market principles' and about 30 to 35 per cent of the state equity has already been privatized. Since 1992 the Government has stressed the importance of privatization to domestic owners.

In 1991-92 foreigners had a dominant role. With the help of new techniques, in 1993 the Hungarians had a major role. Five new privatization methods and their impact on the capital market are analyzed. With E-loans and leasing the transfer of ownership will be limited and slowed because of the debt on owners, and so is the case with MRP (ESOP). In MRPs, the outside investors are usually reluctant to be co-owners (except for the managers). About half of the MRP transactions are hidden MBOs (management buy-out), because the managers have a decisive role. The recompensation bonds issued for earlier owners are used more extensively from mid 1993. The market for these bonds is very liquid. Stock exchange privatization directly increases the role of the capital market. The Pick and the Danubius cases with selling at preferences to small investors were successful.

In early 1993 capital market capitalization reached HUF 300 billion (ten per cent of GDP), of which HUF 50 billion were shares. The yearly turnover is 7 to 15 per cent of capitalization. For the future, securities with acceptable returns and increase in the demand on the capital market are important. For the first, the privatization of public utilities, for the second increased role of small investors and institutional investors would be needed.

The goal of this chapter is to show the interaction between the acceleration of privatization and the development of the capital market. By showing the main privatization techniques, it can be seen that trade investors obtained a more or less dominant position until now, because they depend less on the capital market. On the other hand, from 1993 many companies with a vulnerable liquidity position were privatized as well and their new owners are not abundant in capital.

A larger capital market is greatly needed to increase the demand for privatization as well as to have buyers for those privatized companies which are in need of a capital injection.

Privatization and its main techniques

In the west, capital markets developed after a longer capital accumulation period. In East and Central Europe, the rebuilding of the market economy requires simultaneously both privatization and the development of capital markets.

To minimize the risks of transferring state property to private owners, a strong capital market is needed. In this case new owners can be found relatively easily for those newly privatized enterprises where the new owner has not enough capital needed for profitable investments. In a relatively well-developed capital market with flexible possibilities of ownership transfer, those private persons or institutions who have the adequate means (capital) and capabilities to run the companies will be the 'final' owners. We will analyze the Hungarian privatization from this point of view.

Hungarian privatization from 1990 is firmly based on 'market principles', i.e. state companies must be sold through tendering. Although there were some changes in principles (the role of the bid price somewhat diminished, and domestic owners can acquire state property at conditions more preferable than before), this principle governs the activity of the privatization agencies. In this way, about 20 to 25 per cent of the state equity has already been privatized.

By early 1994 about HUF 180 billion equity was privatized directly by the state property agency (this meant selling more than 600 enterprises and 7,000 shops and restaurants) and about HUF 100 billion indirectly (approving the transfer of state property by the companies to private owners).[1]

In the first two years (1991-92) trade sales dominated with the selling price as the main criteria and the buyers were mainly foreigners (75 per cent of total revenues). When the buyer is a multinational firm, the activity of the Hungarian privatized company, as a subsidiary, is integrated into the firm's overall activity. This means that the Hungarian company's capital requirement is financed by the firm, and not the Hungarian capital market. At middle sized firms in companies having strong domestic ties, this tendency is not so strong.

From 1992 the Government stressed the importance of privatization to domestic owners and decentralization of large state companies. For this aim, new techniques were introduced (mainly increasing privatization demand); decentralization is an important target, moreover commercialization and the tendering procedure is accelerated. With this in 1993 not only the pace of privatization could be accelerated somewhat, but the larger part (60 per cent) of fresh privatization revenue in 1993 came from Hungarian investors.

Table 9.1
Sales according to methods

	1991	1992	1993	1994	Total
MRP:					
number	-	7	128	1	136
revenue*	-	3.13	24.65	0.258	27.03
Leasing:					
number	-	-	9	2	11
revenue*	-	-	3.02	0.67	3.69
E-credit:					
revenue*	0,01	9.07	21.70	1.69	33.47
Compensation ticket:					
revenue*	-	2.2	16.42	0.43	19.05

* in billion HUF
Source: *Privinfo*, no. 5, 1994.

E-credit (existence-credit)

The E-credit at a preferential-interest rate is granted for purchasing state property. Its duration is a maximum of 15 years with a grace period of three years. After the earlier 16 per cent from January 1993, the interest is 3 per cent, and the handling fee of banks 4 per cent. The applicant for the credit must have 2 per cent cash for credits up to HUF five million, and 5 per cent cash beyond that.

Nearly three thousand E-credits took place until early 1993, but 90 per cent were used for pre-privatization (stores) and in only three hundred cases for the purchasing of a state company. The majority of credit holders had enterprises in the past, too. By the end of 1993 the amount of E-credit surpassed HUF 20 billion.

At the moment the four per cent bank fee is quite a problem. For banks this sum is too small, because of high costs and the involved risk, thus the credit banks require a high (usually 150 to 200 per cent) warranty for granting the credit. In order to decrease the risk of banks, the activity of the Credit-guarantee Insurance Company Inc. has been started, but its financial capacities are limited.

Privatization leasing

With this technique those state-owned companies which were earlier unsuccessfully offered for sale can be sold. The leasing company buys the target company, pays a lease fee to the State Property Agency (SPA) for a period of maximum ten years, at a three per cent interest rate. There is no grace period. Until the end of 1993, about 15 companies were sold by this method with a total equity of HUF 1.5 billion. In most of the cases the company management became the new owner.

Employee share ownership program (MRP)

Similarly to the Anglo-Saxon 'ESOP', the employee share ownership program (MRP) was introduced only in the summer of 1992 in Hungary.

In a case of buying out through MRP, as prescribed by law, an MRP-organization must be established to handle the shares during the redemption period and temporarily own the packet. The MRP therefore is an undivided possession during the redemption period, the owner is a legal entity and turns into personal property when this period elapses.

Criterion for the application of this technique is the will of the employees of the organization in question to acquire ownership in the economic company employing them. The precondition of its application is that it be requested by at least 25 per cent of the employees (working there more than six months) in writing. To establish the MRP-organization, 40 per cent of the employees are requested to take part.

The established MRP-organization is a self-governing legal entity. The organization shall be engaged in the buy-out only; it may carry out no other activities. The MRP-organization therefore ceases to exist after repaying the credits drawn. The MRP-organization is a non-profit organization and its activities are legally supervised by the office of public prosecutors.

The financial sources of buying out wealth within the framework of MRP may be credit or deferred-payment. The ratio of own resources shall be two per cent and the term of credit or deferred payment shall be used for paying redemption until fully paid. Interest rate of credits is the same as that of the E-credit in all cases (at present, three per cent). The guardians of state property must therefore use incomes from sales financed this way to lessen debts of the state.

The MRP-organizations established first, and the MRP purchases made can be found primarily in trade. This is no coincidence, since this is the sector that has or had the profitability-volume required for the realization of an MRP. If the situation is further examined, interestingly no large companies or corporations are found on this list. Characteristically their equity is not too large and the success of their activity depends mainly on the firm's expertise and the connecting capital of employees. The most characteristic examples are the foreign trade companies bought through MRP, but the 'Aranypók' store chain can also be included in this category. The interesting point about this latter company is that after selling a substantial number of its shops in the small privatization program the remaining part of the company was bought by the employees themselves.

'Aranypók' is special as, being two years ago a pioneer among the MBOs, nowadays (early 1994) it is planning a modernization of its network by inviting outside private investors to be co-owners. As most of the MBOs are only 1 to 1.5 years old, they could not yet reach this stage.

Another characteristic of the first MRP applications is that they were actually directed at full buy-out. This tendency has continued until today. Almost all MRP cases that are known, that are in process or under preparation, are directed at full buy-out of the state-owned part. The causes are the following:

1   On the one hand many say MRP may be an effective ownership solution for firms which are comprehensible by employees and easily influenceable by their daily work, moreover which do not have too large an equity-value, but are rather profitable. And in these cases it is evident that employees do not wish to share the expected profit.

2   On the other hand the previous practice of the SPA also supports such intentions of investors, since in each case the entire property share possessed by the Agency is offered. The SPA treats the MRP-organization as any other investor, so the MRP-organizations do not dare to submit an application for only a part of the purchasable state-owned property.

3   The third characteristic of these buy-outs, initiated MRPs, is that only such enterprises are included in which capital increase was not necessarily needed for successful operation.

Table 9.2
Revenues of the SPA (billion HUF)

	1991	1992	1993	1994	Jan. 1994
Yields	-	0.94	4.74	2.40	0.08
Selling					
Foreign Currency	0.53	24.61	40.98	25.50	1.11
Forint (cash)	0.14	4.82	17.51	15.30	0.22
Total cash*	0.67	30.37	63.23	43.20	1.41
Credit	-	1.01	9.07	21.70	1.69
Compensation ticket	-	-	2.26	13.00	4.41
Total	0.67	31.38	74.56	77.90	7.51

*   cash and bank transfers
Source: *Privinfo*, no. 10, 1994.

Usually, the first practical difficulty is how the supporters of MRP can win over the decisive and necessary part of employees, moreover what kind of relations are required within the company to initiate the transaction with the hope of success. Already, two important experiences have been obtained in this respect:

1   the majority of employees expect such a program to help them hang on to their workplace, and this has great hidden risks;

2   an MRP initiative without the management and especially against the management has little hope for success. If examined thoroughly, every completed MRP was on the initiative of the management or had background direction from it.

At these companies the employees simultaneously take up two roles: that of the waged laborer and the owner. The advantages of a proprietor's attitude is frequently mentioned.[2] Some troubles can be foreseen. The clash of interests of wage and dividend will inevitably be a source of tension. The issue of layoffs will be another source of conflict.

Private investors are usually reluctant to get equity at enterprises where the MRP has a more than 25 to 30 per cent share of total equity, especially where employees have a veto right over important issues. This means that later they cannot turn to the capital market with success.

Early in 1993 the interest rate of the E-credit used for MRP was decreased radically, which fostered the widespread use of MRP. At the end of 1993, about ten per cent of direct state property privatization was done by MRP (equity of HUF 15 billion). As it is presented in the following point, in about half of the MRP transactions the role of the managers was strong or dominant.

Management buy-out

Management buy-out is a privatization technique present only in practice, without special regulation. This contradiction is closely linked to the ambivalence of the state, on the one hand intending to stimulate manager buy-out and on the other, fearing the previous ruling élite regaining power. The most characteristic feature of today's manager buy-out is that it hides behind employee buy-out, utilizes the allowances existing there, but is not afraid to draw its personal wealth into the venture or offer it as coverage for loans. Where other opportunities exist (such as leasing and E-credit), they are more willing to choose those. In an increasing number of cases the business share intended for purchase serves as coverage for the E-credit taken and credit is paid back from the revenues of the enterprise. This means that there is no fresh capital to the enterprise. On the other hand the managers can bring private investors later to the enterprise. The latter are usually ready to be co-owners with managers. Case studies show that in about half of the MRP-led transfers the management has a strong or dominant role in ownership (see Voszka, 1993 and Karsai, 1993).

Stock exchange

The activity of the Hungarian Stock Exchange is basically restricted by the backwardness of the Hungarian securities market. However, the introduction of Danubius Hotels and Pick Salami Factory shares to the stock exchange was a significant step at the end of 1992.

The HUF 8 billion equity of Danubius was owned by the local government (HUF 1.25 billion) and the SPA (HUF 6.75 billion). The SPA sold two-thirds of its shares: HUF 3.2 billion was offered publicly, HUF 800 million for recompensation bonds and another HUF 400 million for employees with a 50 per cent allowance.

Introduction of Danubius shares to the stock exchange was delayed, since the real estate business is running down on foreign markets and shares do not attract much demand. On the basis of the company's results and future possibilities it was a well-based expectation to introduce Danubius on the stock exchange with success by targeting small domestic investors in the lack of foreign interest.

In the first two weeks of underwriting (1-18 December) only small investors were given a chance, i.e. natural and legal persons intending to purchase less than HUF five million. The days to follow were maintained for large institutional and foreign investors. Following a successful period of underwriting, the security was introduced to the stock exchange on 23 December.

Small investors could obtain their securities by paying and depositing ten per cent on location. For up to 40 per cent of nominal value, the SPA granted a six-month interest-free loan and the sum deductible from the personal income tax served as its coverage; in case of the highest taxation category, the purchaser is returned 40 per cent of the security's price in tax allowance. Thus in this case the allowance is the advance payment of the SPA of this sum by its own credit. The purchaser could acquire a six-year duration E-credit for the remaining 50 per cent of the purchase price, where the security itself is the coverage for this E-credit.

Purchasers had other advantages at their disposal; for every two securities bought, if still possessed after one year, a bonus security was given, and owners of an at least HUF 70,000 value security-package enjoy a 25 per cent discount on Danubius hotel room prices for seven years.

The privatization of the Pick Salami Factory of Szeged was realized in December with stock exchange selling on the initiative and July resolution of the SPA. Originally, the company's management planned this step only after two to three years of reconstruction, but business results and the possibility of strengthening solvency parallel to privatization made the company suitable for the stock exchange in 1992 already.

Privatization through the stock exchange is aimed at allocating shares to small investors and including domestic and foreign financial investors. Nearly 20 per cent of the shares is to be sold against recompensation bonds, of which 5 per cent has already been bought in December, while the remaining 15 per cent is sold biannually in approximately 5 per cent packages depending on the volume of recompensation in circulation. Thirty-five per cent was sold to foreign institutional investors on a closed sale, 18.5 per cent was sold to domestic small investors in an open security issuing at the Budapest Stock Exchange, and another 10.5 per cent was handed to the Commerce and Credit Bank and the Foreign Trade Bank by credit-security exchange; local governments received approximately 4 per cent, while employees received 11 per cent. At the end 5 per cent of shares plus a golden share remains at the State Equity Holding Co. Ltd. Domestic small investors enjoyed a ten per cent allowance, namely purchasers below HUF 300,000 could obtain the securities for 90 per cent of the issue price, providing the share was deposited for one year.

Table 9.3
Foreign investors at SPA, division of investments by country

	No. of Co's	Foreign capital from own capital (billion HUF)	%
Total	356	186.27	100.00
Austria	101	43.18	23.18
Germany	70	34.71	18.63
USA	23	18.73	10.06
Netherlands	13	18.29	9.82
France	34	17.73	9.52
U.K.	28	12.88	6.91
Sweden	8	9.67	5.19
Switzerland	11	8.98	4.82
Belgium	7	5.84	3.14
CIS	13	5.24	2.81
Italy	20	3.89	2.09
Other	28	7.13	3.83

Source: *Privinfo*, no. 5, 1994.

Recompensation bonds

Although recompensation of the earlier (pre-1947) owners by bonds was introduced by the new democratic government in 1991, the large volume of these bonds were issued in 1993 (till end 1993 HUF 70 billion) or will be issued in early 1994. As these bonds can be used only as vouchers to buy state property, their effect on the privatization market has been felt strongly since mid 1993.

By the end of 1993, about HUF 20 billion state property had been sold for recompensation tickets.[3] Usually the recompensed people sell their bonds on the market to brokerage firms who sell it to those domestic investors who are to buy state property. The market is liquid; in the last days of 1993 the daily turnover of recompensation bonds rose to HUF 50 to 100 million (with five trading days a week) at the stock exchange. At the end of 1993 the market value of these bonds stood at about 50 to 60 per cent of their nominal value.

Table 9.4
Sales of businesses (pre-privatization)

	1991	1992	1993	1994	Total
number	4,066	3,571	1,428	174	9,239
revenue*	5.09	6.10	4.55	0.34	16.08

*   billion HUF
Source: *Privinfo*, no. 5, 1994.

An emerging capital market

New instruments of privatization and the capital market

The previously presented methods, easing the terms of privatization loans, are expected to create closed ownership in the circle of the given, primarily small and medium-size companies. Characteristically, banks (and in the case of payment by installments the seller of the SPA) have several instruments even prior to the full repayment of the credit that hinder the selling of companies on the capital market. Banks have mortgage rights and can oblige the prohibition of alienation until the credit is repaid, while in case of other techniques (leasing, payment by installments) quick transfer into possession and ownership only happens if the credits have been repaid. Although the transfer of E-credits to others would be possible, this would probably lead to situations difficult to manage by privatization authorities and banks. At the same time, because of the relatively large preferences (primarily the negative real credit interest rate) the one-time user of the credit is not interested in waiving these preferences, if the new proprietor can also enjoy them. Consequently, more hidden forms of property-transfer will prevail, that is parties interested in the firm, potential capital-owners, will probably purchase property shares or demand an option for the credit it granted.

With the new privatization techniques the government hoped, as of mid 1992, to accelerate the privatization of small and medium size companies for domestic private entrepreneurs, employees and managers. This inevitably results in that it is certainly not the best available proprietor who is selected, rather those who have convincing arguments about utilizing the given capacities in the coming years, and of course who can finance the, usually not too high, loan price. This also means that after some time, especially when new sources (capital, credit) are required for further operation, these companies will be transferred to new owners, who are in the position to mobilize sources for such purposes. Presently, new companies are highly sensitive to external changes for relatively low capital supply and large credits.

A HUF 300 billion ($ 3 billion) transfer of property was envisaged for social security. In mid 1993, HUF 100 billion equity was handed over out of this, in shares of the State Equity Holding Co. Ltd. Negotiations are still underway about the method and subject of the transfer, but at least the partial fulfillment of this goal is expected and thus social security will become one of the determinant players on the capital market.

The Government intends to initiate mass privatization, which mainly means shares marketed primarily on the stock exchange and bought on the account of purchase vouchers provided for citizens. The privatization vouchers of citizens in the value of approximately HUF 100,000 ($ 1,000) could be used on the stock exchange for purchasing state property shares. According to expectations, only minority property shares not exceeding 30 to 40 per cent would be sold for these privatization vouchers. At the same time the majority shares would be acquired by strategic investors.

The effect on the capital market may be a dual one:

1   either masses of small proprietors appear on the stock exchange, greatly increasing the unpredictability and confusion of the stock exchange; or

2   the new small shareholders practice their rights through investment funds. In the case of larger property shares following the business of the given companies, these funds at least partially play a controlling role.

Mass privatization will increase the role of the Hungarian capital market significantly. The present Government's HUF 100 to 150 billion state equity would be enough for the program, as they estimate that at the above terms, 1 to 1.5 million people would take part in the program. These plans are strongly criticized by the opposition parties, and this may be an issue during the spring 1994 election campaign.

New techniques of privatization could be managed more smoothly if the capital market is large and liquid. In the following part we analyze the stock exchange, pinpointing those aspects where changes are necessary.

Capitalization of the Hungarian Stock Exchange grew spectacularly at the end of 1992, but it is still behind the index of similar emerging markets. Capitalization at the end of February 1993 (HUF 289 billion on nominal value, HUF 300 billion on exchange rate value) does not even reach ten per cent of GDP. Out of the total, however, the volume of shares is only about HUF 50 billion (the volume of securities is almost three times as much, that of discount Treasury bonds nearly one and a half times of that).

Compared to capitalization, the turnover of the stock exchange (single) was 15 per cent in 1991 and 7 per cent in 1992. This was caused since, while turnover nearly tripled, capitalization grew more. In 1992 turnover grew to HUF 33.67 billion from HUF 13 billion the year before. Components of the growth are again the securities and Treasury bonds (HUF 12 and 15 billion, respectively), while share sales fell from HUF 12.7 billion in 1991 to HUF 6 billion in 1992; that is below the 1990 level. This shows a drastic distortion of the ratio concerning the Treasury bond: in 1992 it accounted for 44.7 per cent of the stock exchange's turnover, with securities accounting for 36.7 per cent and shares 17.8 per cent. After their introduction one year ago, Treasury bonds gradually crowded out other securities (including state securities) from the market and by the end of the year they accounted for 70 to 90 per cent of the entire turnover.

Another characteristic data is that the increase of turnover took place with the drastic (40 per cent) decline of transactions made, which means a 400 per cent turnover growth per transaction (this in itself is of course a favorable sign).

The further increase of the market concentration of the stock exchange from an already high level is an unfavorable tendency: two-thirds of the share turnover was transacted in shares of three firms. The ownership concentration of firms on the stock exchange is also rather unfavorable from the point of development of the capital market. Moreover, the majority ownership of strategic investors is characteristic mainly of higher ranked firms. This fact inevitably affects the marketing of shares from both the aspect of demand and supply.

Sectional concentration of stock exchange shares is also unfavorable: securities of mainly commercial and service companies are introduced to the stock exchange. With regard to the inadequate diversification of shares and the fact that the concentration comes from exactly the overweight of prosperity-sensitive branches, the economy's crisis resulted in a significant (20 per cent) fall of the stock exchange index. Only three of the 23 companies' shares listed on the stock exchange showed only minor exchange rate decreases and only three showed sizeable improvements.

Although the net savings ratio calculated in the percentage of the total consumer revenue decreased to 11.6 per cent compared to 12.7 per cent a year before, it still greatly exceeds the 1990 data of 5.8 per cent. As regards the entrepreneurial sphere, the net loan-taking position remained, in spite of the fact that during 1992 entrepreneurs were net savers. Deposits of the entrepreneurial sphere grew by HUF 80.2 billion to HUF 462.1 billion during the year. As opposed to this the entrepreneurs' credits equalled HUF 774.2 billion compared to HUF 766.8 billion a year before. The difference of these two sides gives the change of the entrepreneurial sphere's net position, which unusually, but not surprisingly, improved by its 1992 net savings.

Domestic banks and insurance companies are also primarily interested in the purchasing of (state) securities. Beyond the risk/profit factors, they are also motivated by various banking and insurance supervision regulations.

The major obstruction in the development of the stock exchange, in a wider sense the capital market, is the narrowness of demand and supply and its abnormal structure. This has its reasons too. Narrow supply can be traced back to the low number of potential stock exchange issuers in Hungary, because of the recession, structural crisis and inflation. In reality the majority of securities (shares) already introduced to the stock exchange is such.

The result is that stock exchange processes are rather confused, and exchange rates are in no relation to real outputs and expected profits. Certain players can influence exchange rates with one single transaction for speculative goals. And this clearly shows no attraction for issuers of securities, but also scares off investors, both institutional and small private ones.

From this point of view, state securities naturally fall under a different light, because they do not restrict the development of the capital market, at least from the aspect of quantity (in a certain sense they do through the crowding-out effect). It could be seen that the majority volume of stock exchange securities and the turnover of the stock exchange is provided by state securities and this cannot be regarded as advantageous at all. The slow pace and want of issuing long-term state securities does not facilitate the establishment or development of secondary markets.

The appearance of only a small part of savings on the capital market is decisive on the side of demand. As a result of inflation expectations, liquidity and security requirements, the relatively high savings are not primarily attracted by the capital market. The ratio of bank deposits blocked for less than one year is increasing. The other aspect of this problem is that investment funds and other institutional investors play a marginal role on the capital market. Although some investment funds were established in the past year, with success, almost without exception these stayed back from investments in shares. The investment funds concentrate mainly on state securities. What may also have an effect here is that domestic institutions do not have the apparatus suitable for valuing share market processes and individual shares and experienced foreign institutional investors tend to leave the Hungarian capital market.

In the structure of capital market players, the lack of institutional investors is the most striking feature. This could be facilitated by giving equity to social security or mass privatization with the assistance of investment funds.

Conditions for the capital market's development

Positive capital profit and a dividend level exceeding deposit interests. At present in Hungary, the returns of capital are negative on a national economy average and positive only in certain market segments with a rather low positive profit level. Deposit interests exceed the payable dividend-level of the company sphere for the large credit-demand of the budget.

Even the partial correction of these unfavorable conditions is expected only when at least certain major segments of the economy become stable following the recession. First of all this is a question of the market; in the following period the number one task of Hungarian economic policy will be the assurance of a long-term market and not the handling of the short-term market balance.

Securities promising acceptable returns (which can be assured only with the appropriate stability of the economy or the given area). In Hungary the sphere of public utilities and other branches having stable (although somewhat decreasing) markets, even under recession, show acceptable stability when examined on the medium term. This is signalled by the large number of foreign investors in privatization of the food industry and their special interest shown for the privatization of public utilities. By the way growth starts, certain areas linked to consumption and areas connected to branches touched by investment programs (e.g. building material industry) can also become included in this category.

Appropriately functioning institutions. The capital market of Hungary was the first in its region with the opening of the Budapest Stock Exchange in mid 1990. However, if the demand or supply side grows rapidly, significant steps must be taken to increase the technical capacity of the stock exchange.

Adequate capital market demand. The most important elements of this demand are the financial institute sphere, institutional investors, foreigners, households, speculators and brokers. A major step forward could be made within a relatively short period by the appearance of institutional investors in large numbers and in the case of households by mass privatization. In this regard, the transfer of shares for pension funds, investment funds with small investors, and small investors is important. Primarily the shares of public utilities companies could be such securities, together with those of larger companies which have stable markets (one of the instruments to do so is the writing off of a part of the company's loans), and banks. The upswing in the capital market volume would probably be followed by the greater participation of foreign financial funds.

Adequate capital market supply consisting proportionally of securities with stable returns and those financing risky investments, but offering higher profit or capital gains. In Hungary mainly the public utilities companies offer securities with stable returns or, in case of appropriate capital-increase, companies in manufacturing, food industry and commerce with stable domestic markets. Capital gains could be reached by restructuring companies, especially if part of their bad loans are written off.

Division of labor between money and the capital markets, acquisition of sources by the granting of credits and increase of capital with an established, stable institutional system accepted by the players. In spite of the continual decreasing stock exchange index the capital market is in a situation no worse than crediting - since the latter operates in Hungary with an interest gap of around 12 to 15 per cent for the high state draw-offs and the obligation of writing off credits. However, this inhibits the development of the capital market too, because as a result of the low profit and high credit, interest savings do not flow into the economy, neither through banks nor the capital market.

Notes

1. SPA, (1994), *Privinfo Bulletin*, no. 4-5.
2. Private ownership is more efficient than state ownership in: more efficient control of the liquidity; stressing the strategic aspects of corporate governance; and reshaping the capital and asset structure of the company, by increasing the capital and selling the assets not needed.
3. *Privinfo* (1994).

References

Abel, I. and Bonin, J.P. (1991), 'Two approaches to the transformation in Eastern Europe: the big bang versus slow but steady', *Acta Economica*, vol. 43, p. 213.
Charap, J. (1991), 'Transition to a market economy: the case of Czechoslovakia', *European Economic Review*, no. 2-3.
Dhanji, F. (1991), 'Transformation programs: content and sequencing', *American Economic Review*, no. 2.
Frydman, R. and Rapaczinsky, A. (1993), *Privatizáció és vállalatirányítás Közép-Európában*, CEU Press, Budapest.

Giday, A. (1992), 'Close-up on the privatization', *Working Paper Series*, no. 10, Institute for Privatization Studies, Budapest.

Giday, A. (1992), 'Problematic areas of the privatization process at present', *Working Paper Series*, no. 9, Institute for Privatization Studies, Budapest.

Hungarian National Bank (1992), *Annual report of 1991*, HNB, Budapest.

Hungarian National Bank (1993), *Annual report of 1992*, HNB, Budapest.

Hungarian National Bank (1994), *Annual report of 1993*, HNB, Budapest.

Institute for Privatization Studies (1992), *Report on Hungarian privatization 1991*, Budapest.

Institute for Privatization Studies (1993), *Report on Hungarian privatization 1992*, Budapest.

Institute for Privatization Studies (1994), *Report on Hungarian privatization 1993*, Budapest.

Karsai, J. (1993), 'Management buy-out külföldön és itthon', *Külgazdagság*, no. 2.

Kopátsy, S. (1992), 'Mental assets and the development of society', *Working Paper Series*, no. 6, Institute for Privatization Studies, Budapest.

Lindbeck, A. (1971), 'The efficiency of competition and planning' in Kaser, M. and Portes, R. (eds), *Planning and market relations*, Macmillan, London.

Marer, P. (1991), 'Foreign economic liberalization in Hungary and Poland', *American Economic Review*, no. 2.

Matolcsy G. (1991), 'Lábadozásunk évei', *A privatizációs kutatóintézet jelentése* (Our years of convalescence), Report of the Privatization Research Institute, Budapest.

Mellár, T. (1992), 'The economic transition and the economics of transition', *Working Paper Series*, no. 8, Institute for Privatization Studies, Budapest.

Mellár, T. (1992), 'Kereslettöbblet és gazdasági növekedés' (Excess demand and economic growth), *Bankszemle*.

Newbery, D.M. (1991), 'Reform in Hungary' (Sequencing and privatization)", *European Economic Review*, no. 2-3.

Sári-Simkó, A. (1992), 'The new Hungarian acts on privatization', *Working Paper Series*, no. 11, Institute for Privatization Studies, Budapest.

State Property Agency (1993), 'Privatization in Hungary', *Privinfo*, various issues, Budapest.

Szapáry G. (1993), *Macroeconomic issues of transition in Central and Eastern Europe*, Presentation at a conference organized by the Atlantic Research and Publication and the Budapest University of Economic Sciences, Budapest.

Vanicsek, M. (1993), 'Privatization in the balance', *Working Paper Series*, no. 12, Institute for Privatization Studies, Budapest.

Voszka, E. (1993), *Self privatization as an alternative to centralized selling or to free allocation of state equity,* Center for International Private Enterprise, Budapest.

10 Perspectives on financial systems and enterprise efficiency in the transition economies of Central and Southeastern Europe

Stavros Thomadakis[1]

Introduction

The focus of the analysis in this chapter is the promotion of productive enterprise in the economies-in-transition. This focus follows not only from the general need to reshape the productive system in ex-socialist economies, but also from the specific observation that production has suffered large declines during the initial stages of transition in every economy in the region of Eastern-Central Europe. The main purpose of analysis is consequently defined by the requirement to institute mechanisms which will lead to an attainment of increased efficiency by productive enterprise. There are several distinct aspects to efficiency. One is technical efficiency, in the sense usually understood by economists as the location of productive activities on an available production possibility frontier, rather than inside it. A second is organizational efficiency, understood chiefly as the ability of the enterprise as organization to respond to change and pressure, and to attain smooth managerial operation without intra-organizational conflicts, including conflicts between various stakeholders in the enterprise. A final aspect, which is more complex because it encompasses the previous two, is market-related efficiency: the ability of the enterprise to absorb and act upon market signals in a fashion that will allow its survival, its renewal and attainment of new levels of productivity, and ultimately, its growth.

An important element of market-related efficiency is the relation of the firm to its capital market environment. This relation is the central concern of the chapters in this volume. The capital market can encompass a variety of mechanisms and institutions ranging from stock markets to various types of banks. It is ultimately this complex of mechanisms and institutions, which we call the capital market, that controls the flow of capital into or out of a particular business venture; this flow is regulated by assessments and expectations which capital market participants come to hold about the prospects and capabilities of any given firm. A firm enjoying high prospects becomes a pole of attraction for capital market funds. A firm with low prospects finds itself increasingly rejected in the quest for market funds. It is by such a process that the capital market is understood to furnish 'discipline' to firms: it rewards efficient managers and firms but penalizes inefficient ones.

Usually, in developed market economies, one or the other aspect of efficiency furnishes the primary roadblock to enterprise success or survival. For example, industrial firms in long-protected industries may suffer from serious inefficiency of the technical kind, and must therefore focus their restructuring effort on cost-cutting and retooling efforts. Or, for another example, firms overburdened with debt are faced with the primary task of restructuring claims between debtholders and residual claimants in order to survive, if they still own assets of value. In the case of transition economies, however, the task of restructuring is much more complex because all aspects of efficiency goals are pressing at the same time, and obstacles arise on all fronts: technical obsolescence, state ownership, and economic reflexes reared in bureaucratic bargaining rather than competitive market signals; all have to be tackled and changed in some fashion simultaneously. In actuality all problems of such complexity eventually succumb to the limitations of human choice and political feasibility, which must simplify the problems and start from somewhere, even if it is admitted that simple approaches might be imperfect. The emphasis on privatization of state enterprises reflects precisely such a course. Although the complex goal of restructuring must be achieved ultimately, one can start with a well-defined albeit one-dimensional action: shifting ownership away from the state and into the hands of private interests. The shift of ownership to private interests opens up significant new problems associated with company governance. Decentralized decisions about the utilization and valorization of capital must bear the burden of restructuring, of technical renewal, of the attainment of international competitivity, and ultimately of the economic development of their country under conditions of free markets and democratic rights. This is already a very tall set of desiderata. It should be approached under the operation of the best possible institutional arrangement which will smooth rather than magnify inter-company conflicts, inefficiency or self-inflicted uncertainty. The link between company management and capital market arrangements can be very valuable in facilitating these tasks and ensuring a high probability of success.

The remainder of this chapter is organized in six sections. Section two brings forth several common features from the experiences of transition to date. These features inform the remainder of the analysis. Section three discusses several issues arising from the transitional 'mixed economy': public-owned and private firms must coexist but this coexistence must be prevented from degenerating into subservience or systematic subsidization of one kind of firm by the other. The conditions for successful generation and utilization of domestic financial savings are discussed in section four. Section five links the discussion and analysis in previous sections with the evolutionary potential for institutions in the financial sector. Section six offers summary comments and conclusion.

Some common features of the 'state of transition'

The process of transition in the economies of Eastern-Central Europe has now been underway for several years. The economic literature regarding the transition has also accumulated at a relatively impressive rate, albeit with emphasis on macroeconomic aspects of the process. In any case, the collective experience of transition in the countries of East-Central Europe is long enough to furnish valid notions about the peculiarities of a 'state of transition'. Several important features of this 'state of transition' appear to qualify as common characteristics of the economies undergoing the process; such common characteristics must be incorporated in any credible analysis, or policy design, related to company management and capital markets in the transition economies. The requisite listing of features draws upon the country studies by Abel and Prander (1994), Dumitriu and Nicolaescu (1994), Giday and Sári-Simkó (1994), Hrncír (1994), Mujzel (1994), Panov (1994) as well as the comparative chapter by Bilsen (1994), included in this volume.

Privatization is a time-consuming process

It appears that, due to capital shortages, organizational constraints, and political misgivings against selling off too many large firms at the same time, effective privatization is feasible at a relatively limited scale and at a limited rate over time. It should be noted that mass privatization schemes such as those encountered in wide distributions of coupons to the population at large (viz. Czech Republic, Romania, or Poland) represent only a nominal form of privatization, and are not really the end of the privatization process but rather only its beginning. The creation of an effective block of private interests which exercises control over large state-owned firms is still a problem awaiting its solution in the case of the mass privatizations which have taken place or are under way. This solution may be a time-consuming process as well. Therefore, there will exist for a considerable time span - five or even ten years perhaps - a peculiar mixed economy. Privatized firms and newly-born private enterprises will co-exist with firms which remain in the state sector, and also with firms whose ownership status remains in limbo somewhere between the public and the private sector. The mechanisms which push for efficiency in those firms which remain in the state sector or, even more significantly, those whose ownership status is unclear, must not operate at the expense of the enterprises in the private sector or vice versa.

Capital market development faces systemic constraints

Active open capital markets normally have three functions. They act as a funding mechanism, they underpin the allocation of risks, and they also function as a valuation mechanism.[2] These three functions are highly intertwined. The funding function is, for example, highly linked to the formation of capital structures and the consequent governance structures of firms since it is funding choices that shape the classes of capital stakeholders in an enterprise.[3] This same funding function will prove paramount at the current state of the Eastern-Central European economies, because of their large needs for investment capital. Via the funding function 'outsiders' (e.g. financiers, banks, private portfolio investors) will sooner or later be exercising constraints on firms' decisions and strategies.

To fulfill that function and to act as an efficient funding mechanism, (i.e. as a mechanism which actually raises capital under market conditions), capital markets in the transition economies must resolve problems on both sides of their operations. As a mechanism for the collection of savings they must overcome a number of obstacles which limit the potential for financial savings. Low personal incomes and low present consumption levels combined with higher living standard aspirations, inflation and institutional risks, represent primary obstacles. At a second pass, the predicted needs for internally-financed investment in nascent private sectors will act as a secondary obstacle for the rapid increase in financial saving by the business sector of society. On the other hand, traditions of high private saving within the household, and some natural precautionary preference for private saving under circumstances of widespread uncertainty about individual and collective economic outcomes could represent factors favorable to the development of financial saving, provided that the capital market can offer adequate financial products for the satisfaction of these needs. The setup of capital market institutional arrangements, and the pursuit of requisite policies must take into account these facets of potential saving.

On the other hand, the capital market as a mechanism for the allocation of savings to investing units must be able to discriminate among uses in at least two respects. On one hand, it must be able to distinguish between bona fide, well-intentioned entrepreneurs and impostors.[4] In economies where no reputations have been built up, and no implicit trust in business relations between parties has had a chance to accumulate, this distinction is no easy matter. On the other hand, agents in the capital market must also be able to attribute correctly the projects undertaken by bona fide entrepreneurs into risk classes and charge a risk premium accordingly. This is a function which also requires experience and ability to perform correct risk-evaluations of good projects. To fulfill these functions, capital market agents must have skills in collecting, processing and evaluating information. In emerging markets these skills are usually in short supply.[5] In the ex-socialist countries, the shortage will probably prove even more severe given the lack of experience with markets generally, and the virtual non-existence of commonly accepted accounting rules and conventions which normally form the basic code underlying valuation logic.

Problems of consistency in policies and fiscal integrity of the state must be factored in the analysis

Aside from economic-technical uncertainties which characterize the function of all market economies, the economies in transition are subject to exceptional uncertainties which are political and institutional. These arise basically from the process of reconstruction of the role of the state and the broader public sector, which forms part of the overall transition. There are at least two important elements in these uncertainties. On one hand, there is the problem of continued political support for economic reform policies, within the context of the democratic process in the countries of East-Central Europe. Several authors have warned that the economic pain of reforms, especially those administered as 'shock therapy', endanger the consistent application of reforms over time by creating social protest that is bound to be expressed in political debates and in electoral preferences for parties which profess a slowdown or even a reversal of reforms.[6] Recent political developments in several transition countries, the length of time and political debate about privatization in virtually all East-Central European countries testify to the political complications of structural reform. The end result is uncertainty about the time-path of reforms which can lead, in turn, to postponement of investment and other long-term decisions, thereby prolonging recessionary trends and reinforcing the uncertainty of the time-path of the economy itself.

The second element which feeds political-institutional uncertainty is the ambiguity surrounding the prospective fiscal integrity of the states in East-Central Europe, during and after the transition. There is a clear danger that the transition to market-type economies will go too far in denuding the state of economic powers and will engender weak and deficit-prone fiscal structures. A confluence of factors engenders this ambiguity. Firstly, the state is shedding many of its assets via mass privatizations, thereby disassociating itself from potential or actual sources of revenue. On the other hand, the state is under pressure to overtake some liabilities, as for example in the case of socialization of old outstanding debts of SOEs. Secondly, the buildup of effective tax systems in the context of market-type economies is bound to be a time consuming and politically complicated process itself. Thirdly, the pressure for social relief from the pains of transition will inevitably lead democratic governments to increase public expenditure for consumption. Fourthly, the pressure for renovation of infrastructures, and the provision of new public goods, such as environmental restitutions, is real and necessary for underpinning sustained private investment. Naturally, state resources will also be economized by means of massive subsidy removal. However, the probability of continued deficits and of the accumulation of new state debt is nontrivial.[7] Uncertainty about the fiscal integrity of the state can prove deleterious to both the reform process and the efficient functioning of the private sector; phenomena such as the crowding out of private securities in the capital market and/or inflationary pressures from monetary deficit financing engender the risks of instability in capital markets and a climate of speculative behaviors which is inimical to long-term investment.

Given these principal features of the 'state of transition', the analysis which follows seeks to pose issues of financial system design and of company management arrangements which will clarify how best to achieve efficiencies during the transition, and how to safeguard the conditions for high performance after the transition is over. It is understood, naturally, that the premise of the analysis is that the 'state of transition' is not a temporary episode but a process with considerable time length and breadth.

Allocational issues in the transitional 'mixed economy'

The transitional 'mixed economy' will be composed for a considerable period of time of privately-owned and state-owned enterprise. The latter will in principle include enterprises which are more or less in limbo in terms of final ownership. These enterprises will naturally not operate in mutual isolation. On the contrary, they will interact and become associated in many ways; for example through production linkages, technical linkages, financial linkages, or even strategic linkages. Several issues arise about rules that will govern the relations between these types of enterprise. The three issues which are addressed relate to transitoriness in the status of public sector firms, to corporate taxation policies, and to enforcement of financial contracts between firms.

Transitoriness and distortions

The transitional 'mixed economy' differs from a steady-state mixed economy in the sense that in the former, enterprises remain in the public sector not out of strategic or social choice, but rather because of the presence of obstacles to their rapid privatization. The emphasis on insufficient demand for the purchase of whole or fractions of enterprises, which is brought forth by Giday and Sári-Simkó (1994) in the case of Hungary, underlines precisely the nature of one of the main obstacles to privatization. The necessarily transitional character of public sector firms

262

may then create problems both for their own operation and for the operation of private firms which come into contact with them. Those who currently manage and control public sector firms will typically anticipate the termination of their control over those firms and will consequently choose policies and actions which are rational for short-run horizons but irrational over long-run horizons. Thus, 'short-termism' may be induced by the very prospect of a privatization whose completion, however, is not well defined on a temporal scale. Generally, this type of behavior leads to an emphasis on trading activities, but to a corresponding deemphasis of investment activities which boost production potential or quality. Even more dangerously, those who manage public sector firms will in the circumstances commit them to actions which insure their own (the managers') well-being but which damage the firm's productive capability all the more easily. In other words they may become completely self-serving 'agents' in a context where the 'principal' (the state) is a self-declared lame duck and does not exercise any meaningful monitoring of their actions.[8] As Mujzel (1994) aptly observes, the decline in large firm profitability in Poland cannot be entirely attributed to environmental factors. It must be also explained by internal systemic reasons. The fuzziness of ownership status and the short-termism of induced management incentives are the most likely basis of these systemic reasons.

An inactive 'principal' can be the cause of a fundamental asymmetry between the firm's assets and liabilities. Thus, consider the case of a state enterprise which has 'actives' (assets) which are primarily in the form of physical capital, raw materials and inventories of finished goods; and also has 'passives' (liabilities and equity) made up of debts to banks, other firms, ownership claims etc.[9] In the special condition of transition in the economy as a whole, and transitoriness in the ownership status of the firm, there arises significant asymmetry between 'actives' and 'passives' in the sense that whereas the former are physical entities, and therefore reasonably well-defined, the latter are contractual entities, and therefore fuzzy by definition. The fuzziness is not limited only to ownership claims. It also extends to most outstanding debts. These debts were incurred in many cases for relatively recent investment expenditures of SOEs, and are held by state banks as nonperforming assets. The final resolution of the problem of nonperforming debts is itself a policy issue and the fuzziness in debt claims will persist until and unless this issue is resolved.[10]

The asymmetry in the character of 'actives' and 'passives' is valuational. The value of physical assets is in effect more readily ascertainable and less prone to errors or miscalculations than that of contractual liabilities. The status of the latter is subject to political decisions, bargaining, and general institutional arrangements to be put in place in the future. In short, assets are marketable and liabilities are not. In general, for firms whose ownership status is transitory, the 'passives' are less meaningful than the 'actives'.[11] This asymmetry can create incentives to liquidate those 'actives' which can be easily valued, before clearing up the status of the 'passives'. In the presence of effective monitoring by the owner-state such liquidations may simply be a rational way to undertake step-by-step privatizations and to reduce the scale of public sector firms. Such steps would presumably use the proceeds of liquidation to rearrange or adjust the remaining 'passives' (e.g. pay off some debts). However, in the absence of effective monitoring by the owner-state, the process may be undertaken by the actual managers (or other parties in day-to-day control of enterprise actions) for their own benefit. Such benefits can materialize in a great variety of forms ranging from cash kickbacks on business transactions of the public sector firm to acquisition of shares in private business activity. The appearance of this result will become even more probable if it can be sufficiently obscured under the pretext of crisis, unclear accounting practices, or 'necessary restructuring' of the state enterprise. In this way however, the productive ability and the efficiency potential of the enterprise will be damaged.

In sum, the lack of effective monitoring by the 'transitory' owner, who is the state and/or state-owned banks, enhances short-term and selfish behaviors on the part of de facto managers of these enterprises. The reflection of this in the firm's structure is that liabilities appear to be much more fuzzy than assets. This asymmetry then tends to be corrected not by the rectification of liabilities but, in effect, by the erosion of the assets. The institution of effective monitoring is of course a problem in itself. If quick and effective privatization is not forthcoming, as is apparently the case in all East-Central European economies, then there are only two intermediate routes for policy. One is the institutionalization of overseeing mechanisms (e.g. holding companies, regulators, supervisory boards). The other is the devolution of effective control to existing managers and workers. The first type of arrangement is complicated, requires skills which are in short supply, could engender politicking, and may eventually pose more problems in the form of a need to regulate the overseers. It could, however, prove effective in cases where strategic choices for future restructuring are relatively clear, and where there is easy measurability of results. Furthermore, this solution could also prove more efficient in terms of choosing the timing of an effective shift of ownership to the private sector.[12]

The devolution of effective control to existing managers and workers is a solution which could take the form of leasing - as has already been the case in Poland, for example - or the form of a medium-term management contract with options to retain fractional ownership at the end of the contract. These (or other) forms of devolution of control would in many cases align actual day-to-day control which managers and/or workers exercise anyway with control of longer-term decisions and outcomes. The advantages of this solution would be to lengthen the time horizon of those who exercise de facto control, thereby reducing the risks of asset liquidation or dissipation. The disadvantages are that strategic choices may be systematically biased towards immediate benefits as opposed to investment commitments, and to the use of labor intensive techniques. The disadvantages could also extend to a bias against a future choice of privatization which would transfer control out of the hands of management and workers. In part, however, the latter problem can be resolved by the use of long-term options as part of the endowment of manager/worker owners or lessees.

It may not be possible to generalize ex ante on the applicability of one or the other solution, and therefore it may be desirable to allow institutional degrees of freedom for both of them. The 'organizational pluralism' suggested by Mujzel (1994) in the case of Poland is probably of more general applicability in the remaining East-Central European economies. This is especially fitting in the cases of the more backward economies of Bulgaria and Romania, where the institutional vacuum is more evident. It probably forms the basis of what Panov (1994) argues are 'spontaneous markets'. It also is very obviously the cause of the great lags between legislation and effective change at the micro-economic level which is underlined by Dumitriu and Nicolaescu (1994) in the case of Romania. The application of 'organizational pluralism' can also be seen as a chance for institutional evolutionary processes to work themselves out.

Corporate taxation and differential burdens

Corporate taxation presents important issues in the context of a transitional 'mixed economy'. It is widely argued and commonly agreed that ex-socialist economies must construct tax systems practically from scratch.[13] Specifically in the area of corporate taxation it is necessary to achieve the replacement of implicit tax revenue from state enterprises, which formed the foundation of central planning practices of the past, by explicit tax revenue based on profit or value-added taxes.[14] The question of tax systems lies outside the scope of this study and the associated research. However, the construction of a tax system has direct implications both for capital markets and for corporate capital structures. More specifically with respect to the transitional mixed economy, the process of constructing a new tax system for corporate entities can arguably lead to differential treatment between types of enterprise. Scenarios which make that likely are briefly discussed here.

It is a fair statement that part of the economic transition for ex-socialist societies devolves upon a reorganization of the revenue functions of the state. However, the ability of the state to organize effective tax enforcement, and the speed with which effective enforcement can be mounted are matters of conjecture.[15] If state weakness and political vicissitude translate into weak tax enforcement, several episodes of history suggest the inevitable recourse of states to an inflation tax, through the medium of monetary finance of deficits.[16] In such a case, it is worth inquiring how even or uneven the incidence of an inflation tax can be within the business sector. In fact, the incidence of the inflation tax may differ between state-owned and private enterprises, for a variety of reasons. For example, the import-intensity of production or the export intensity of sales, the relative ease of extracting indexed wage contracts from state enterprise managers, the capacity to hold large inventories of physical goods, the ability to borrow in money-fixed terms, and finally the amount of outstanding nominal debt, may constitute good reasons for differences in the inflation tax burdens of state-owned and private firms. It is clear from that listing that the tax-incidence bias of inflation could work in either direction as an empirical matter.

State weakness may not be so acute as to lead to inflation. It may be sufficiently present, however, to lead to differential willingness or capability for tax enforcement in the two spheres of the economy. This can be the outcome of a confluence of political and technical factors. Normally, one would expect that it is easier, and politically more convenient within the politics of transition, to enforce taxes upon state-owned enterprise than upon private enterprise, especially when the latter is newly established, hotly desired, and shrouded in an aura of vigor and purity juxtaposed to the perception of a corrupt public sector. Lack of information, non-uniform accounting standards and the moral force of 'independence from a long oppressive state' may all work against effective tax enforcement on the side of private firms during the transition. On the other hand, state-owned enterprises may fend off their own tax burdens either by utilization of their accumulated losses and their accumulated debts from the past, or by sheer ability to maneuver and to apply pressure within political institutions. Furthermore, pressure for quick sell-offs to private interests may also encourage the provision of incentive packages in the form of future tax forgiveness, which may end up being self-defeating from the fiscal point of view. The vicissitudes of tax reform and tax enforcement in nascent market economies may in fact require that extremely simple forms of taxation of broad application are primarily used. For example a sales tax or a VAT would be the least susceptible to systematic differentiation of tax treatment between types of enterprise.

In any case, equal tax treatment between enterprises of the public and of the private sector is required, as Mujzel (1994) directly argues and Panov (1994) implies, not only on grounds of equity. Any systematic differential tax treatment between the two sectors of the economy, irrespective of the actual mechanisms by which it comes about, will create phenomena of tax arbitrage. These will be tendencies for taxable gains to show up in the low-tax area of the economy and for tax deductible losses or expenses to gravitate towards the high-tax area. Tax arbitrage will, in other words, create a tendency to transfer visible/measurable profits from the high-tax to the low-tax spheres. If the high-tax sphere is the state sector, then this implies that state firms will tend to understate profits and private firms will tend to overstate them.

The opportunity for tax arbitrage can generate incentives on the part of state-enterprise managers to manipulate the relations of their enterprise with private firms. For example, by selling underpriced outputs or purchasing overpriced inputs from private firms, state-owned enterprises may reduce their tax liability. In effect, state-owned enterprise managers would undertake such actions in order to shift potential government revenue to private profit, provided that they can obtain a share out of private profit better than out of government revenue. Eventually, this tax arbitrage will affect not only the way activities are priced but also the organization of activities itself.[17] Thus, it is plausible that high return projects or activities, even if conceived within the processes of a state-owned firm, will become preferably located in the private sector, where they can generate higher private net benefits given a certain level of gross benefits.[18] Another inevitable, and not so desirable, effect of this process, however, will be a tendency for the decline of tax revenue from the corporate sector of the economy. This would be in effect the case if tax avoidance on the part of the private sector led to widespread phenomena equivalent to the 'underground economy' encountered in several Western countries.[19]

One important distortion which can arise from tax arbitrage phenomena is that visible/measurable profitability of private firms will not be reflective of true profitability. Mujzel (1994) suggests this as a possibility for the explanation of declines in measured profitability in Polish enterprises. If this comes to pass as a systematic feature of the business sector, both efficiency measures and entry signals in the private sector will be distorted with attendant misallocations. Additionally, widespread practice of tax arbitrage will also lead to buildup of strong resistances to accounting reform and the imposition of stringent standards for financial reporting, within the corporate sectors of transition economies.

Financial enforcement

The enforcement of financial contracts is another area in which, just as in the case of taxes, differential capabilities between private and public sector enterprises will have important distortionary effects in the transitional economy. Financial relations between state-owned firms and private enterprise can arise both as a result of trade between them (e.g. via the granting of merchandise credit) and through the sale or exchange of securities between them. The latter, however, is probably a subject directly related to privatization of state-owned enterprise, and will not be taken up here. The former is more germane to the present discussion.

Practically all the authors of country studies in this volume point out the growth of inter-enterprise indebtedness during transition. Hrncír (1994) especially pinpoints the large quantitative significance of the phenomenon and its wide ranging implications. Inasmuch as inter-enterprise debt is forced upon lenders by the inability of borrowers to pay for their purchases, its growth in the period of economic transition reflects two fundamental features of the countries' economic-institutional conditions. On the one hand, it describes a reflex on the part of firms to circumvent restrictive monetary policies, imposed through the tightening of credit at the level of central or commercial banks. On the other hand, it also shows that this reflex, which can be observed in many economies under similar conditions of 'credit crunch', can be accommodated widely due to the absence of credible enforcement rules, which is in itself an institutional failing. Thus, complaints about the ineffectiveness of monetary policy are in essence directed at the lack of financial enforcement mechanisms.[20]

The growth of inter-enterprise indebtedness normally follows the lines of merchandise supply. Suppliers who offer credit to their customers may 'net it' against their own suppliers. However, in any such system there are net borrowers and net lenders. Net lenders may in fact be borrowers from banks, or they may lend by running down their cash positions. If trade credit is not honored in time, it is net lenders who bear the cost. In a system of interconnections by means of enterprise debt, the occurrence of a bankruptcy will snowball from net borrowers to net lenders. Or, short of bankruptcy, the tightening of budget constraints in the form of reduced subsidies or bank credits places a higher burden on net lenders - who risk seeing their credits go bad - than on net borrowers, who have an option to forcibly extend their debts. If it is therefore the case that the distribution of net lender or net borrower roles is not random in the economy, but rather represents a systematic outcome of trade linkages, then the firms which find themselves in net lender positions will be discriminated against by tight monetary policies. If these are private sector firms, for example, they may be forced to reduction or cessation of operations due to insolvency. If, on the other hand, they are state sector firms, they may either reduce their operations due to insolvency, or they may use their condition to petition the government or monetary authorities to soften their credit constraint. Thus, the location and type of net lenders in an inter-enterprise system of debt makes a difference as to the ultimate effects of restrictive monetary policies. Unfortunately, the country studies in this volume and elsewhere have not been able to furnish information in such detail. The most efficient resolution of the problem, however, would require such information.

It is also a practical possibility that different firms have different abilities to enforce their lending agreements. For example large firms may be more able to collect from customers than small firms, or state sector firms may be better able to throw their weight around and force payment from their debtors.[21] However, as in all financial contracting, different abilities to enforce do not exclusively determine the outcome. Different willingness to pay must also be taken into account. It is not an unknown phenomenon (viz. several Greek public enterprises in the late 1970s and 1980s) to have private firms systematically accumulate arrears in their debts to public enterprises, because the latter are either too slow or too badly organized to push for timely collection. Furthermore, if state-owned firms are known to be more likely to obtain a softening of their credit constraint, they can become natural targets of payment delays and defaults.

The evidence on the accumulation of inter-enterprise debts which is presented by Abel and Prander (1994) and Hrncír (1994) in this volume shows a clear quantitative explosion of the phenomenon in the period of transition. As Hrncír (1994) has aptly described it, this represents widespread 'credit contract failures', and it leads to inevitable erosion of the integrity and the faith in all debt contracts. The absence of bankruptcy laws and practice, of ex ante risk premia and ex post reputational penalties for the abrogation of debt contracts has probably contributed to this failure. The tightness of monetary policy, the recession, and the sudden withdrawal of state subsidies in some cases have also contributed to it by constraining tightly the liquidity of firms, as Abel and Prander (1994) argue. It is not possible to attribute shares of the responsibility to these two categories of causes. However, it is clear that the restitution of the faith and the integrity of debt contracts is essential for the entire process of both the rationalization of external financing of firms and the ability of capital market agents to contribute positively to the governance of enterprise.

More generally, the attainment of efficient operation in the context of the transitional 'mixed economy' can be impeded by the transitional (temporary, nonfixed) features of the enterprises remaining in the state sector. It can also be sidetracked by incentives which arise from transitional tax and/or financial enforcement arrangements (or absence of arrangements) which influence and distort the initiatives of private firms, orienting them towards short-term benefits.

Aspects of financial system design

The importance of financial saving to economies in transition cannot be overstated. In general, the existence and operation of financial saving is coterminous with the existence and operation of capital markets. Financial saving implies the existence of a supplier (saver) and a user of savings (borrower) who are different entities and who, under specific contractual and institutional arrangements, come into contact in the market to exchange present for future monetary resources. Financial saving is that portion of an individual's, a household's, or a firm's saving which is turned into monetary/financial assets, i.e. into claims for payment in the future. It is, in aggregate, the supply of currently available resources to the capital market, broadly defined, which encompasses both arms-length markets for securities, and intermediated markets for claims, which are generated by banks or other financial institutions.[22]

Many commentators on the economies in transition have stated that the development of a capital market, broadly defined, is an essential requirement for sustainable development of a healthy private sector in these economies. Two interconnected primary reasons are usually invoked. These are related to the fundamental functions of a capital market. The first is the need for a mechanism which transfers resources from net savers to net investors. The importance of this function must be underlined in any context where the distribution of savings differs from the distribution of either investment opportunities or entrepreneurial skills, or both. The second is the need for that mechanism to operate as a device for transferring resources to the best investments, or of not renewing the transfer of funds to bad investments, or, finally, of ranking investments and charging their finance with differential risk premia. This is the part of the capital market function which produces 'discipline' for private firms and their managers.[23]

The effective function of capital markets is always underpinned by microstructural conditions which support successful outcomes. In any financial transaction which involves future promises for repayment, availability and quality of information is one such microstructural condition which enables claimant parties to ascertain the future ability to pay of their contracting counterparties. Means of establishing credibility furnish another microstructural condition which enables claimants to ascertain, in addition to ability, the willingness to pay of their contracting counterparties. Finally, the existence of benefits (i.e. profits) to be reaped by the capital market agents who act as intermediaries in financial transaction is another necessary condition, since there are transaction costs to be covered.[24] In the following paragraphs of this section these issues are considered in turn, and in relation to the specific problems of transition economies, as they were discussed above.

Issues of information supply and of credibility

The effectiveness of capital markets in producing 'discipline' depends on information about the future prospects of any investment. As Stiglitz (1985), and Greenwald and Stiglitz (1986), have argued persuasively, information asymmetries are more pronounced in capital markets (as compared to goods markets) and lead to either market failures or rationing, or to a requirement for significant government regulations. These outcomes can moderate substantially the force of 'discipline' that a free capital market can mete out successfully. Most of the problems in the efficient function of capital markets arise from inability to distinguish, ex ante, between good and bad projects; or in addition, from inability to distinguish, ex post, whether a good (or bad) outcome was the result of performance or luck.

There are several reasons for which informational problems are heightened in transition economies. The first is investor-manager inexperience. The formulation of a private business project requires some prior knowledge of market conditions, of technique, of organizational requirements for successful completion and operation of a project; in newly formed private sectors, it may be the case that even the investor-manager himself does not have good command of these aspects of his own project and does not have the means to accurately assess the quality of his investment, let alone communicate effectively his assessment to others in the capital market. Thus, the primary supply of information can be flawed, at its very source. In addition, inexperience can also be a problem for other entities along the chain of financial transactions, for example, entities which assess project quality at arm's-length, e.g. banks, other financial intermediaries, market analysts or capital market agents. This magnifies the 'noise' surrounding true assessments of project quality and attempts to rank investment projects correctly. Finally, primary savers who act as financial investors in capital markets are most likely to suffer from inexperience with the result that they may become easy prey to mistakes, misinformation, or 'fads' which are not based on company fundamentals. In those circumstances, demand and pricing of financial instruments will be inefficient.[25] The capital market, instead of producing 'discipline', may simply end up producing opportunities for profiteering and speculative schemes.

The second cause of informational problems is institutional uncertainty. In section two we discussed problems of uncertainty which arise from the possible time-inconsistency of economic policies and from the uncertainty about the future fiscal integrity of the state in transition economies. Institutional uncertainty has microeconomic aspects also, and these concern us here. Thus, there is uncertainty about the value of preexisting financial claims, about potential new tax liabilities, and about mechanisms for distribution of residual wealth upon a firm's bankruptcy or liquidation. All these areas represent matters which must and will be settled by policy decisions and possibly by legislation sooner or later. Their delayed or their partial resolution creates a cloud of probabilities where there could instead be certainties. This institutional uncertainty affects adversely the wealth positions of primary financial savers and of financial intermediaries, and consequently biases their decisions against financial saving.

The third cause is the absence or relative weakness of regulations and overseeing activities which protect primary financial investors from fraud, insider profiteering, and other harmful financial schemes which transfer wealth among investors, especially in the area of risky (or speculative) investments. Again, this is not simply a question of absence of legal structure but also, and more so, a question of absence of skills.

Under the conditions of these heightened informational problems, it is premature and perhaps even risky to start up capital market functions by instituting arm's-length markets (security markets) as the primary space of financial saving transactions. The performance of these markets would be too erratic, too subject to misinformation and manipulation, and too sensitive to speculative bubble phenomena; in short, such a market would suffer from serious credibility problems and these would in effect prevent it from becoming an effective mechanism of capital transmission and control. It would be fair to predict in fact that in markets functioning with such features the probable end result is that, after a few bad experiences, these markets would dry up as a source of new funds for a long time. In the end, their premature establishment and their inadequate legal and informational infrastructure could prove to be a negative rather than a positive contribution to the development of permanent financial saving structures and potential. The history of several speculative booms and busts in western markets confirms this danger. A distinction is called for here, however. The preceding comments are really directed to security markets in which stocks or share type securities are traded. Debt securities are far simpler and involve far fewer informational requirements than stock securities. This is a result of the fact that debt securities constitute financial contracts whose performance is far more well defined ex ante than company shares. Thus, debt securities may be better suited as effective instruments in an informationally noisy environment where creditworthiness and credibility are hard to establish. Furthermore, debt securities can also play a significant role in furnishing external control to firms since monitoring is also in the interest of debtholders.[26] The problems associated with dispersed ownership which occurs in equity markets can also occur in debt markets, however, if bondholders (and creditors in general) are too many. Dispersed ownership creates incentives for 'free riding' and is likely to lead to suboptimal monitoring as every security holder hopes to free ride on the monitoring performed by others. Furthermore, in the case of many creditors there are costs of coordination.[27] Two implications seem to follow. First, if security markets are to be developed as a matter of policy measures, bond markets should probably take priority, or at least should not be ignored.[28] Second, the general disadvantages of broad based security markets in terms of offering effective external control seem to lead to a preference for banks and structures of financial intermediation as the primary space of financial savings transactions.

Issues of banking and financial intermediation

To organize financial saving transactions around the operation of financial intermediaries has obvious advantages, but it also presents problems. Keeping in mind what was said in the previous paragraphs about the dangers of credibility gaps in financial markets, it is plausible to suggest that needed buildups of credibility are easier to achieve around a structure of financial institutions which can also more easily become the subject of both government guarantees and overseeing, if necessary. The major advantages of financial intermediaries - e.g. commercial banks or variants of investment-holding companies - are that not only can they be monitored more easily but also that they can learn, or receive know-how, to conduct assessments and to exercise monitoring in their turn, and do so more quickly than a large collectivity of independent but inexperienced investors. In an environment as informationally noisy and fluid as the transition economies, monitoring is a quasi-public good of major significance, and its importance is heightened precisely because of the higher relative difficulties of ex-ante distinctions of good from bad investments and of risk assessments. Taking for example the case of wide dispersion of ownership claims, such as the distribution of coupons or shares in state enterprise in the Czech or Romanian schemes, there arises an obvious need for monitoring by an agent who will 'champion' the interests of small stakeholders. This is a function which goes beyond the usual regulatory protection of investors from fraud or manipulation of prices, and extends to the assurance that small minority stakeholders will not be penalized or expropriated by large ones. A 'champion' is not simply a constraint formulator. It is an active agent of small stakeholder interests, who can endow the function of securities transactions with credibility.[29] Arguably, such a champion must be an institution which combines some of the following features:

1   It can be, to some extent, publicly-owned and monitored inasmuch as it produces a type of monitoring which creates positive externalities in the conditions of developing markets.

2   It can be, to some extent, constrained to align its actions to small stakeholders by allowing the latter to own options against its portfolio.

3   It can be, to some extent, performing an intermediation function in the sense that it may issue different types of claims than the ones it holds.

In extremis, such an institution could be simply a holding bank issuing sure deposit claims to supplement its own capital, enjoying a public guarantee of its liabilities, and holding a portfolio composed primarily of debt but also of some equity obligation of industrial firms.

Another very significant source of advantages associated with financial intermediaries is their ability, as known again from western market experience, to reap economies of scope from the combination of transactions and payments services with credit screening and allocation decisions. If it is assumed that East-Central European economies will achieve monetary stabilization from the macroeconomic policies of transition, the development of payments systems, and emerging needs for payments services will be rapid and variegated in view of the large new population of private firms which will take up erstwhile public economic space. Banking organizations which are able to fill this need will then be able to also reap the associated informational advantages for credit

273

screening and allocation. These benefits will translate to increased profits, will cover costs of transactions, and will help to maintain high incentives for the banks to expand the field over which they exercise these functions. In this fashion, capital market development can become the outcome of strategies of participating organizations rather than just the hoped for result of policy and legislation imposed from above.

Finally, it has been argued effectively that banks are capable of supporting much more effectively long-term finance because of their ability to form, cultivate and maintain long-term relationships with their borrowers.[30] Such relationships enable both more effective monitoring on the part of banks and more successful long-term policies on the part of non-financial firm managers.[31] Long-term relationships can of course become the basis for implicit contracts and for relations of trust, which ultimately offer an avoidance of costly problems of agency and can counteract the harmful aspects of competitive markets for hostile takeovers and corporate control.[32]

The main disadvantage of a financial system which will rely primarily on banking organizations in the countries in transition is the historic burden of bad loans and unperformative portfolios of state-owned banks. Experts appear to agree that this problem must be faced radically and quickly either with a massive write-off of debts or with a separation along the lines of 'good banks - bad banks' strategies.[33] Such separations usually imply that the 'good bank' part will do business with the nascent and dynamic private sector whereas the 'bad bank' will in some way remain as an encumbrance of the state. In fact, such a strategy may be effective only if the inter-enterprise debt tangle does not transcend the boundaries of public and private sectors. If this is not the case, and if private and state-owned firms are linked in the inter-enterprise debt system, then an effective separation of 'good bank' from 'bad bank' can only occur after debt write-offs or restructuring in the form of 'socialization', e.g. the exchange of enterprise debt with government debt. These are not insuperable problems, however, in a context of monetary stability and of financial transactions protected from outbursts of speculative activity.

In the case, finally, where besides breakaways from the original socialist monobank, there is also some new entry of banking institutions, an important point must be noted. 'Old' banks (originating from the monobank) will carry many nonperforming loans on their books. They will be under constant pressure to lend more to old and nonperforming clients in hopes of ultimately recovering some part of their sunk investments. On the other hand, with respect to new clients, they will have an incentive to be very selective so as not to magnify their problems. They will also attempt to charge higher prices for their loans and services in order to defray some of the costs of their nonperforming portfolio. On the other hand the 'new' banks may find it easier to work with new and riskier clients who are probably screened out by the 'old' banks. This type of story appears to fit pretty well the facts and the analysis presented in the very perceptive chapter by Hrncír (1994) in this volume; specifically high spreads between deposit and loan rates and more risky undertakings by 'new' banks, as Hrncír reports for the Czech Republic, seem to indicate that the 'new' banks are entering at the high risk end of the lending business, whereas the 'old' banks have captured the low risk end but are burdened of course with most nonperforming loans. The obvious danger of the situation is that the 'new' banks may also become quickly burdened with bad loans, in which case the reputation of the whole banking sector will suffer. It would therefore appear that simple prudential regulations (e.g. capital adequacy requirements) must become rapidly deployed.

In lieu of conclusions: the evolutionary potential of financial systems and their links to nonfinancial enterprise governance

Developed capital markets operate to allocate funds, to structure maturities and to distribute risks. These functions are carried out by a complex of arm's-length markets and financial institutions, by transactions in formal or informal markets, and by millions of implicit decisions of primary savers about the form of their savings, financial or real. When faced with the problem of designing a financial system almost from scratch, it is easy to succumb to the temptation to provide a full set of arrangements for carrying out all these complex functions. However, this can lead to a type of 'institutional overdetermination'. That is, it can lead to a complex of arrangements which do not allow room for learning, for experimentation, and for evolution. Yet, in the face of nascent private sectors whose final shape cannot be predicted, in the face of indeterminacy about the final economic role and limits of the state, in the face of uncertainty about the future rates of growth and directions of development, 'institutional overdetermination' can create unnecessary costs and rigidities of future adjustments.

The debate on choice between a bank-based system and a securities market-based one cannot ignore that where these systems exist, they arose from specific conditions of private sector development and of private-public sector relations. Furthermore, this debate cannot also ignore that even within classes of systems (e.g. bank - based ones) there arise very significant differences among countries and tradition (viz. Japanese versus French banking, for example), although there are of course common features also. Again, this underscores the fact that the shape of institutional arrangements in the financial sphere should be responsive to the course of private sector development, the distribution of productivity and savings potential among economic units, and the structure and distribution of entrepreneurial skills and investment opportunities in the economy.

One way to avoid 'institutional overdetermination' is to begin from very simple capital market functions: for example, allocation of capital at the lowest scale of the risk-spectrum and at relatively short maturities. This would be equivalent to orienting (or even restricting) banking operations largely to low risk loans (high quality, collateralized loans, for example) and low risk liabilities. Or, in securities markets or interbank markets, it would be equivalent to a major orientation to short or medium-term fixed-income securities. There are several advantages to such a policy, whereas its disadvantages do not appear to be debilitating for economies in transition.[34] The main advantages are:

1 the creation of economy-wide and accessible risk-free financial assets which will act as anchors to personal or enterprise financial strategies, and which will also allow a clear formulation of a primary market interest rate;

2 the protection of banking institutions early in the transition from major valuation mistakes, whose likelihood grows with the riskiness of their credit allocations;

3 the supply, in the spirit of McKinnon's (1992) view, of monetary asset substitutes to real asset accumulation by enterprises with respect to their needs in working capital;

4 the pressure on entrepreneurs to rely on self-finance for their more risky undertakings, so as to maximize their incentives for screening and risk assessment and for learning to economize capital;

5 once credible and sound financial institutions are consolidated in a low risk market, the degrees of freedom for new arrangements which can accommodate higher risk capital allocations will probably be increased.

Notes

1. I thank the participants of the Prague ACE Conference for helpful comments. Special thanks are due to Marvin Jackson for many valuable suggestions.
2. Stiglitz (1992) offers an exhaustive list of capital market functions (pp. 162-3).
3. See in this connection Williamson (1988), who clarifies within a transactions cost framework the fundamental relations between attributes such as asset specificity and the preference for alternative forms of external financing.
4. This is the basis for the famous adverse selection problem in which inability to distinguish between 'good' and 'bad' projects leads to perverse pricing of risk and ends up penalizing the 'good' projects. See Akerloff (1970).
5. See Kitchen (1992), Ojo (1992).
6. On the centrality of political issues see Sachs (1993), especially Chapter 1, for a view of the most articulate proponent of 'shock therapy'; see also Przeworski (1992) for an opposing point of view, which argues even more strongly about the centrality of political opinion.

7. See the very enlightening contributions of Tanzi (1993), on the relation of public finances, financial markets and their respective reform in transition economies.

8. The classic exposition of divergence of incentives between non-owner managers and passive owners, and the consequent emergence of 'agency costs' is given in Jensen and Meckling (1976).

9. I am borrowing terminology used in Jackson (1992). Naturally, contractual entities can also be present on the side of firm 'actives', as in the obvious case of inter-firm debt claims. Yet, speaking of industrial firms, it is a fair presumption that most contractual entities arise on the side of 'passives'.

10. I am indebted to Marvin Jackson for pointing out the need to distinguish between fuzzy and non-fuzzy debt obligations.

11. Stiglitz discusses the uncertainties of valuing bank assets in the transition economies. Bank assets are state firm liabilities by and large. The fuzzy valuation of these liabilities is in effect what Stiglitz refers to as 'risks of undervaluation' of banks in ex-socialist economies. See Stiglitz (1992), pp. 171-2.

12. The experience of privatization of 'problematic' firms in Greece is suggestive. The formation of an organization for the restructuring of enterprises (OAS in Greek) provided an opportunity for appropriate timing of privatization. On the other hand, it also entangled privatization efforts with political calculations and power struggles between the government and large public-sector banks which owned many 'problematic' firm securities. The German experience with the Treuhandanstalt is a more centralized version of the same principle. See in this respect the comparative exposition in Carlin and Mayer (1992).

13. For specific contributions in the area of tax policy see the volume edited by Tanzi (1993).

14. McKinnon (1992), makes a persuasive case for the need to establish a modern tax system. His policy recommendation is, in sum, that old style profits taxation be maintained for state sector industrial firms but that an EC type VAT be established for newly private enterprises.

15. Tanzi (1993), suggests that the actual horizon for the stabilization of the institutional aspects of the tax system may well be ten years.

16. See for example Dornbusch (1988), who points out the importance of seigniorage for inability to correct deficits by taxation may be either structural - e.g. large populations of small firms or self-employed who can easily evade tax - or ideological, as in the case of conservative tax reduction choices.

17. The argument here parallels the well known arguments and expositions of 'transfer pricing' used by multinational firms which operate across tax jurisdictions. There is a voluminous literature on the subject. For classic expositions see Lall (1973) and Lessard (1979).

18. This distortionary outcome may be seen as justified if it ultimately boosts the dynamism of the private sector. Or it may be seen as a form of implicit privatization. The issue ultimately relates to the deeper question of the role of the state revenues and their alternative uses. E.g. the development of a private sector also requires the supply of public goods, infrastructure etc.

19. A voluminous literature has accumulated on the 'informal sector' or the 'underground economy'. See Tanzi (1982), Cowell (1990).

20. Hrncír (1991) suggests that money supply figures (for former Czechoslovakia) must be corrected: "... for the change in volume of financial obligations not honored at due time ..." (p. 321). This redefined money supply would in essence be a concept corresponding to an institutional arrangement of weak financial enforcement.

21. Kornai (1990) states that: "The prospects for recovering debt are much brighter for an influential state-owned firm than for an unaided private firm" (p. 141). This presumably reflects experience in Hungary, and definitely refers to conditions before the change in regime. As a general statement however, it may or may not be true, as it assumes that debt collection motivation is as strong in state-owned firms as in private firms.

22. For classic views on the significance of financial savings and capital markets see Goldsmith (1969), McKinnon (1973), Shaw (1973). For a specific perspective on motive factors for security markets development see Kumar and Tsetsekos (1992).

23. See Stiglitz (1992).

24. It is noted here that intermediaries are not present in bank-mediated markets only. They are also present in security markets as investment bankers, market makers, brokers, dealers and generally agents who typically intercede between original seller and ultimate buyer. The difference of bank-type intermediation is that banks may differentiate their claims from their liabilities, whereas in direct security markets intermediaries do not change the nature of the securities they handle.

25. Of the large literature on security market efficiency I note the classic exposition in Fama (1970). I also note recent contradictions to efficiency theory on the basis of findings of 'investor sentiment' in the stock market. See Poterba and Summers (1988).

26. See in this connection Aghion and Bolton (1988) and Mayer (1990). For an interesting perspective on debt securities as payment for privatized enterprises and of debt payments as a substitute for tax payments see Bolton and Roland (1992).

27. See Mayer, ibid., pp. 323-4.

28. See Senbet (1992), for an argument in the case of developing economies and emerging markets.

29. I have argued this point for the case of emerging capital markets in less developed countries, where variants of development-investment banks can play the role of 'champion'. Without a 'champion' it may be impossible to establish confidence in arm's-length security markets. Even with a 'champion' the establishment of confidence may take time. See Thomadakis (1992).

30. See Corbett and Mayer (1992), for an application of this argument to the need for a bank-oriented financial system in transition economies.

31. For evidence on the better performance of firms with large institutional links via bank borrowing and bank presence on the board of directors of German firms see Cable (1985). For the antipodes, see an empirical confirmation of how securities markets induce 'short-termism' in Miles (1993).

32. On the importance of implicit contracts and relations of trust between parties to a variety of transactions see Kester (1992), who presents hypotheses and evidence on

industrial groups as systems of effective governance. For a related but more financially oriented rationale of the formation of industrial groups with a bank at the center see Thomadakis (1992).

33. See the contribution of Calvo and Frankel (1991), who rightly give high priority to the task of 'cleaning' the balance sheets. See also Thorne (1992), on policy recommendations for Bulgarian financial reform. Thorne suggests that banking for the private and the public sector should be segregated at the institutional level.

34. The disadvantages from a restricted supply of high-risk finance relate mostly to the high relative cost imposed on genuinely high risk ventures, e.g. technological innovations. This is a limited problem in transitional economies where, most probably, technological innovations will attract foreign financing in fully-owned subsidiary form or joint ventures, for example.

References

Abel, I. and Prander, K. (1994), 'Impediments to financial restructuring of Hungarian enterprises', this volume.

Aghion, P. and Bolton, P. (1988), *An incomplete contract approach to bankruptcy and the financial structure of the firm*, mimeo.

Akerloff, G. (1970), 'The market for lemons: quality of uncertainty and the market mechanism', *Quarterly Journal of Economics*, vol. 84, no. 3.

Bilsen, V. (1994), 'Privatization, company management and performance; a comparative study of privatization methods in the Czech Republic, Hungary, Poland and Slovakia', this volume.

Bolton, P. and Roland, G. (1992), 'Privatization policies in Central and Eastern Europe', *Economic Policy*, October.

Cable, J. (1985), 'Capital market information and industrial performance: the role of West German banks', *The Economic Journal*, March.

Calvo, G. and Frenkel, J. (1991), 'Credit markets, credibility and economic transformation', *Journal of Economic Perspectives*, Fall.

Carlin, W. and Mayer, C. (1992), 'Restructuring enterprises in Eastern Europe', *Economic Policy*, October.

Corbett, J. and Mayer, C. (1991), 'Financial reform in Eastern Europe: progress with the wrong model', *Oxford Review of Economic Policy*, no. 4.

Cowell, F.A. (1990), *Cheating the government: the economics of evasion*, MIT Press, Cambridge MA.

Dornbusch, R. (1988), 'The European monetary system, the dollar and the yen' in Giavazzi, Micossi, Miller (eds), *The European monetary system*, Cambridge University Press, Cambridge.

Dumitriu, I. and Nicolaescu, T. (1994), 'Delayed privatization, financial development and management in Romania during the transition', this volume.

Fama, E. (1970), 'Efficient capital markets: a review of theory and empirical work', *Journal of Finance*, May.

Giday, A. and Sári-Simkó, A. (1993), 'Privatization and capital markets', this volume.

Goldsmith, R. (1969), *Financial structure and development*, Yale University Press, New Haven.

Greenwald, B. and Stiglitz, J. (1986), 'Externalities in economies with imperfect information and incomplete markets', *Quarterly Journal of Economics*, no. 1, March.

Hrncír, M. (1991), 'Money and credit in the transition of the Czechoslovak economy' in Siebert, H. (ed.), *The transformation of socialist economies*.

Hrncír, M. (1994), 'Financial intermediation and company management: case of the Czech Republic', this volume.

Jackson, M. (1992), 'Practical equity and efficiency issues in the privatization of large-scale enterprise, *Working Paper*, LICOS, Katholieke Universiteit Leuven.

Jackson, M. (1994), 'Property rights, company organization and governance in the transition, this volume.

Jensen, M. and Meckling, W. (1976), 'Theory of the firm: managerial behavior, agency costs and capital structure', *Journal of Financial Economics*, vol. 3, October.

Kester, W.C. (1992), 'Industrial groups as systems of contractual governance', *Oxford Review of Economic Policy*, no. 3.

Kitchen, R. (1992), 'Venture capital: is it appropriate for developing countries?' in Fischer, K. and Papaioannou, G. (eds), *Business finance in less developed capital markets*. Greenwood Press, Westport, Connecticut.

Kornai, J. (1990), *The road to a free economy*, Norton, New York.

Kumar, P.C. and Tsetsekos, G. (1992), 'Securities market development and economic growth' in Fischer, K. and Papaioannou (eds), *Business finance in less developed capital markets*, Greenwood Press, Westport, Connecticut.

Lall, S. (1973), 'Transfer pricing by multinational manufacturing firms', *Oxford Bulletin of Economics and Statistics*, August.

Lampe, J. (1992), 'Introduction' in Lampe J. (ed.), *Creating capital markets in Eastern Europe*, Woodrow Wilson Center Press, Washington.

Lessard, D. (1979), 'Transfer prices, taxes and financial markets: implications of internal financial transfers within the multinational firm' in Hawkins, R. (ed.), *The economic effects of multinational corporations*, Jai Press, Greenwich, Connecticut.

LICOS (1992), *Interim report: company management and capital market development in the transition*, Leuven.

McKinnon, R. (1973), *Money and capital in economic development*, Brookings Institute, Washington.

McKinnon, R. (1991), 'Financial control in the transition from classical socialism to a market economy', *Journal of Economic Perspectives*, Fall.

McKinnon, R. (1992), 'Taxation, money and credit in a liberalizing socialist economy' in Clague and Rausser (eds), *The emergence of market economies in Eastern Europe*, Blackwell, Oxford.

Mayer, C. (1990), 'Financial systems, corporate finance, and economic development' in Hubbard, G. (ed.), *Information, capital markets and investment*, University of Chicago, Chicago.

Miles, D. (1993), 'Testing for short termism in the U.K. Stock Market', *The Economic Journal*, November.

Mujzel, J. (1994), 'State-owned enterprises in transition: prospects amidst crisis', this volume.

Ojo, A.T. (1992), 'Problems of orthodox formal capital markets in developing countries with particular reference to Nigeria' in Fischer, K. and Papaioannou, G. (eds), *Business finance in less developed capital markets*, Greenwood Press, Westport, Connecticut.

Panov, O. (1994), 'Delayed privatization and financial development in Bulgaria', this volume.

Poterba, J. and Summers, L. (1988), 'Mean Reversion in stock prices: evidence and implications', *Journal of Financial Economics*.

Przeworski, A. (1992), 'Economic reforms, public opinion and political institutions: Poland in the Eastern European perspective' in Pereira, L.C.B., Maraval, J.M., and Przeworski, A. (eds), *Economic reforms in new democracies*, Cambridge University Press, Cambridge.

Sachs, J. (1993), *Poland's jump to the market economy*, MIT Press, Cambridge MA.

Senbet, L. (1992), 'Research directions: a theoretical framework' in Fischer, K. and Papaioannou (eds), *Business finance in less developed capital markets*, Greenwood Press, Westport, Connecticut.

Shaw, E.S. (1973), *Financial deepening in economic development*, Oxford University Press, New York.

Stiglitz, J. (1985), 'Credit markets and the control of capital', *Journal of Money, Credit, and Banking*, vol. 17.

Stiglitz, J. (1992), 'The design of financial systems for newly emerging democracies in Eastern Europe' in Clague and Rausser, ibid.

Tanzi, V. (1982), *The underground economy in the United States and abroad*, Health and Co., Massachusetts.

Tanzi, V. (1993), 'Financial markets and public finance in the transformation process' in Tanzi, V. (ed.), *Transition to market*, IMF, Washington.

Thomadakis, S. (1992), 'Notes on firm governance and growth: the theory of the firm in less developed capital markets' in Fischer, K. and Papaioannou, G. (eds), *Business finance in less developed capital markets*, Greenwood Press, Westport, Connecticut.

Thorne, A. (1992), 'Reforming financial systems in Eastern Europe: the case of Bulgaria' in Lampe, J. (ed.), ibid.

Williamson, O. (1988), 'Corporate finance and corporate governance', *Journal of Finance*, July.

4l+8